King's Highways & Steam Trains

A Ford Mainline to the Branchlines of Ontario in July 1957

Engine 5578 on train 174 at Shallow Lake, August 1958 — W. MacLaren Wilson photo

VOL. 1

Ian Wilson

To John with best wishes, Ian Wilson Nov. 8/16

IanWilsonAuthor.com

Nonfiction by Ian Wilson

Steam at Allandale
To Stratford Under Steam
Steam Over Palmerston
Steam Through London
Steam to the Niagara Frontier
Steam in Northern Ontario
Steam Scenes of Allandale
Steam Echoes of Hamilton
Steam Memories of Lindsay
Steam at Washago

Fiction by Ian Wilson

The Secret of the Old Swing Bridge
The King's Puzzle

In memory of Mom and Dad and their '56 Ford

Wasaga Beach, August 1958 — W. MacLaren Wilson photo

— Mary Lou Wilson photo

Train 168 at
St. Marys,
July 22, 1958.

—John Rehor photo

A Ford Mainline to the Branchlines of Ontario in July 1957, Vol. 1

King's Highways & Steam Trains

ANGUS SITS IN THE DRIVER'S SEAT OF A PARKED 1956 FORD. REACHING INTO his wallet, he pulls out a Motor Vehicles Operator's License. Old Eddie prepared that for him. The Province of Ontario issued it, ostensibly, in 1957. *Angus Wolfe, male, age 21, height 5'-7".* That's him, all right. He tucks the paper back, next to a wad of 1950s Canadian banknotes.

It's time to get his bearings. From his vantage point at the side of a curving street, he gazes through the windshield. Perhaps 50 yards away is the rear of a railway station, constructed of brown brick. The track is beyond, on the far face of the building. Hanging from its eaves on the right end is a blue sign. Its white letters spell CANADIAN NATIONAL EXPRESS. Directly below, attached to the end wall of the structure, is a smaller sign. NO PARKING. FOR EXPRESS AND ROYAL MAIL ONLY. And just now, steam locomotive 5594 is pulling in from the right, bell ringing. A mail & express car, an open-window coach and a refrigerator car trail the engine.

Gripping the steering wheel, Angus feels the vibrations of the idling '56 Ford through his fingers. It looks as if the harmonic frequency set up by the inline six cylinder engine of Old Eddie's car has worked. It has transported him back almost six decades in time, to July 1957. Again.

Where exactly am I? Angus wonders. There's some kind of landmark out the passenger window. It towers above everything in the vicinity. Leaning over the front seat, he glances upwards. It's on the far side of a street that rises to cross the railway track on a wooden bridge. He fixes his eyes on the structure. It's a water tank, sitting atop an octagonal base made of limestone. There's a slogan painted on its body. ST. MARYS... THE TOWN WORTH LIVING IN.

Angus reaches for a 1957 Imperial Oil road map beside him. Those produced by that retailer were—are—the only ones depicting railway lines. It's among the provisions Old Eddie handed him this morning. *Happy Motoring... starts at the Esso sign.* The index puts St. Marys in

Spring 1984.
—Mary Lou Wilson photo

square J-11. He opens the map in front of him. It blocks out the sun like those folding windshield screens people put in parked cars. There's St. Marys. Roughly halfway between London and Stratford. A good three hours drive from my hometown, he figures. The Canadian National mainline through here runs north-south. That means it's early morning, by the light angle. There's also a branchline running west from here to Sarnia. I can't be parked beside that one, though, judging by the sun. There's even a Canadian Pacific line from Ingersoll. What a decade the 1950s was—is. Railway lines everywhere. He refolds the map and lays it on the seat.

That water tank out the passenger side looks familiar. Of course it does. Angus has been here before. It wasn't for long. He was with his parents, years ago. Or several decades in the future, as the case may be. On that overcast May afternoon, he and his dad stood on the station platform. The very one where engine 5594 and her three cars are now. He had a hardcover book in his hands, then. It detailed the comings and goings of trains on a 1950s summer day, in words and pictures. He positioned himself from the vantage point of the photographers while reading the text. On that visit, he was trying to place himself back in the Age of Steam.

Now Angus cranks down the driver's window of the '56 Ford. A wave of jubilation rushes through his veins. What he had imagined years ago has come to pass. This *is* the Age of Steam.

TORONTO — GUELPH — KITCHENER — STRATFORD — LONDON — GODERICH

Read Down — Read Up

39 Daily	37 Ex. Sun.	141 Sun.	175 Ex. Sun.	11 Ex.Sa. & Sun.	171 Ex. Sun.	111 Sat.	29 Ex. Sun.	137 Sun.	27 Ex. Sun.	Miles	TABLE 100 Eastern Time	10 Daily	28 Ex. Sun.	172 Ex. Sun.	34 Ex. Sun.	36 Ex. Sun.	138 Sun.	170 Ex. Sun.	40 Daily
P.M.	P.M.	P.M.	P.M.	P.M.		P.M.	A.M.	A.M.	A.M.			A.M.	A.M.	A.M.	P.M.	P.M.	P.M.		P.M.
10.30	6.20	5.50	5.35	4.20	..	12.10	8.10	7.50	6.55	0	Lv Toronto, Ont. (Tables 80, 81) Ar	7.00	9.50	11.10	7.35	8.35	9.15	..	11.00
10.39	6.30	6.00	5.44	4.30	..	12.20	8.20	8.00	7.06	2.4	Parkdale	6.50	9.42	11.01	7.24	8.27	9.06	..	10.52
10.49	6.40	6.06	5.53	4.38	..	12.28	8.27	8.08	7.15	5.0	West Toronto	6.39	9.35	10.54	7.16	8.20	8.58	..	10.45
f10.56	6.47	..	6.02	4.47	..	12.36	..	f8.16	7.23	8.6	Weston	6.32	..	10.46	..	8.12	8.50	..	..
11.07	7.02	..	..	..	..	12.47	..	f8.25	7.40	15.3	Malton	6.16	..	10.36	..	8.01	8.40	..	..
11.19	7.16	6.28	6.22	5.07	..	12.58	8.52	8.36	7.58	21.3	Brampton	6.06	9.14	10.26	6.47	7.50	8.30	..	10.23
11.31	7.26	6.39	6.35	5.18	..	1.10	9.02	f8.50	8.16	29.3	Ar Georgetown (Table 90) Lv	5.53	9.02	10.13	6.37	7.30	8.13	..	10.11
11.31	7.35	6.39	6.41	5.18	..	1.10	9.06	f8.50	8.16		Lv Georgetown Ar	5.53	9.02	10.13	6.37	7.30	8.13	..	10.11
..	a7.40	..	..	..	..	..	..	..	..	32.5	Limehouse	..	..	f10.05	..	..	f8.06	..	..
11.44	7.49	..	f6.55	5.31	..	1.22	..	f9.03	8.30	35.6	Acton	5.40	..	10.00	..	7.13	8.01	..	..
..	f7.56	..	..	..	..	f1.30	..	f9.10	8.39	41.2	Rockwood	..	..	9.50	..	7.04	f7.51	..	..
12.10	8.13	7.05	7.25	5.57	..	1.45	9.40	9.25	9.00	48.8	Ar Guelph (Table 104) Lv	5.04	8.31	9.40	6.10	6.53	7.40	..	9.45
12.10	8.13	7.05	P.M.	5.57	..	1.45	9.40	9.25	9.00		Lv Guelph Ar	5.04	8.31	A.M.	6.10	6.40	7.40	..	9.45
..	..	..	..	..	..	..	..	..	..	49.8	Guelph Jct.	..	..	..	..	6.34	..	..	..
..	f8.29	..	To	..	..	..	d9.56	f9.41	..	58.4	Breslau	..	8.10	From	c5.46	6.21	f7.21	..	..
12.45	8.45	7.27	Owen	6.15	..	2.12	10.04	9.50	9.30	62.7	Ar Kitchener (Table 109) Lv	4.37	8.04	Owen	5.40	6.15	7.14	..	9.20
12.45	8.45	7.27	Sound	6.25	..	2.12	10.04	9.50	9.30		Lv (Waterloo 2 mi.) Ar	4.37	8.04	Sound	5.40	6.15	7.14	..	9.20
..	f8.54	..	Table	..	..	..	a10.14	f10.02	9.41	69.2	Petersburg	..	7.50	Table	c5.25	5.44	f7.00	..	..
..	9.02	..	104	..	..	f2.25	a10.18	f10.06	9.47	72.5	Baden	..	7.41	104	c5.20	5.36	6.54	..	..
..	9.10	..	..	a6.41	..	2.30	a10.21	f10.09	9.53	75.0	New Hamburg	f4.13	7.38		c5.16	5.32	6.50	..	n8.58
..	f9.20	..	..	..	..	..	a10.28	f10.18	10.03	82.1	Shakespeare	..	7.28	168	c5.06	5.21	f6.40	..	..
1.18	9.30	8.01	..	7.00	P.M.	2.48	10.38	10.30	10.15	88.6	Ar Stratford Lv	3.52	7.17	A.M.	4.55	5.10	6.30	P.M.	8.40
1.30	↓	8.04	..	7.10	6.00	2.55	10.45	A.M.	A.M.		Lv (Tables 102, 111) Ar	3.47	↑	7.08	↑	P.M.	P.M.	7.05	8.35
..	↓	..	..	..	f6.10	..	..	↓	..	93.9	St Pauls	..	↑	Ex. Sun. 6.54	↑	..	..	f6.54	..
f1.46	↓	8.23	..	7.25	6.23	3.12	11.03	↓	..	99.7	St. Marys	3.28	↑	6.45	↑	..	..	6.46	8.17
..	↓	..	..	..	f6.33	..	..	↓	..	105.7	Kellys	..	↑	6.25	↑	..	..	f6.33	
..	↓	f8.37	..	..	6.42	3.27	11.18	35	..	110.7	Thorndale	..	↑	6.17	↑	..	..	6.27	..
2.20	↓	8.55	..	8.00	7.05	3.45	11.40	Ex.Sun.	..	121.4	Ar London (Table 81, 103) Lv	2.55	↑	6.00	↑	..	..	6.10	7.45
A.M.	↓	P.M.		P.M.	P.M.	P.M.	A.M.	A.M.				A.M.		A.M.				P.M.	P.M.
	9.45							10.45	..	88.6	Lv Stratford Ar	..	7.05	..	4.40				
..	f9.57							10.57	..	93.7	Sebringville	..	6.54	..	4.27				
..	10.08							11.12		101.8	Mitchell	..	6.40	..	4.12				
..	f10.19							11.22	..	106.9	Dublin	..	6.31	..	4.00				
..	10.31							11.36	..	113.4	Seaforth	..	6.19	..	3.46				
..	10.45							11.58	..	121.7	Clinton	..	6.04	..	3.25				
..	11.10							12.20	..	134.5	Ar Goderich Lv	..	5.40	..	3.00				
	P.M.							P.M.					A.M.		P.M.				

Note—Tickets reading via Can. Nat. Rys. between Hamilton or Dundas and points beyond Kitchener will be honored by Canada Coach Lines between Hamilton or Dundas and Kitchener.

It's easy to send a
PREPAID TICKET
Your local agent will give you the full particulars.

He touches his fingers to the metal roof. Ouch—hot. It's summer all right, just like in the book he was reading on that other visit. But the morning sunlight of St. Marys in July 1957 won't last long. Any moment, he and the time-travelling car will be whisked back to the March snows around Old Eddie's garage at Washago. Back to Angus's regular life in small-town Ontario, a couple of hours drive north of Toronto. Back to the second decade of the 21st century.

Curious about the train that just arrived, Angus reaches for a booklet under the Imperial Oil map. It's a Canadian National Railways timetable, in effect from April 28 to October 26, 1957. The cover illustration depicts two diesel locomotives snaking a passenger train through the mountains. *Air conditioning, foam-rubber seats and picture windows assure you of a restful trip in our modern coaches.* He flips through the pages to the index. There are two columns of abbreviated Saints. He runs his finger down the names and comes to St. Marys. It's on Table 100. He leafs back and finds it. That table covers trains in the Toronto-Guelph-Kitchener-Stratford-London-Goderich corridor. Given that it's early morning, there's only one train locomotive 5594 could have pulled in—number 168. It runs from London to Palmerston via Stratford, daily except Sunday. And that warning sign on the station isn't kidding—a group of men are already loading and unloading sacks of mail and express shipments on the platform.

Grasping the ignition key in his left hand, Angus turns the engine off. The fluttering sound and vibrations cease. This is quite something, he ponders, being back in July 1957.

Angus thinks about Eddie, the guy who got him here. Locals sometimes call him Old Eddie. No wonder—he's well past 80 years of age. Angus encountered him for the first time when he was a toddler, hunting for frogs and turtles on his grandmother's property. That's at Washago, a short drive north of his hometown of Orillia. Eddie's dad ran a White Rose gasoline station there, beginning in the 1920s. Later, the business passed from father to son. It was—and still is—located on what had been Highway 11 through the village, near the Trent Severn Waterway. Eddie was on hand nine summers ago, on the day beside the canal when Angus had encountered one Amanda Webb. Funny how the old fellow kept turning up in his life. Eddie later helped Angus and Amanda solve some family mysteries relating to steam-era trains.

Eddie has spent decades restoring pre-1960 autos and trucks. One of those was a green 1950 GMC pickup for Amanda's grandfather, Big Jim. Eddie works on the vehicles in the service bay of his old White Rose station. The place has otherwise been closed for business since the pumps were removed in the early 1960s. That was after the bypass was put in and the old

—author photo

swing bridge over the canal was removed. Those improvements marooned a section of old highway and killed any resident trade.

The car in which Angus sits, Eddie's latest restoration, is special. It's a 1956 Ford Mainline in Bermuda Blue. Angus remembers learning about it last fall. Often, when he's in Washago to visit Amanda, he drops in to see what Eddie's working on. Angus loves anything to do with the 1950s. Stepping into Eddie's garage is like a trip back in time itself. The day after last Halloween, Angus ducked into the service bay of the old White Rose station.

"Whatcha got there, Eddie?"

Eddie popped his head out from under a raised hood. "Hiya, Angus." He tapped the vehicle's front left fender, which rattled in place. "It's a '56 Ford."

On that day, Angus begged to differ. What Eddie had been working on was plainly a derelict. The body of the car, what was left of it, had long since lost the battle with rust. All four tires were flat. The rubber fragments were so brittle they would have to be carved off the rims. Wooden blocks under the axles kept the carcass off the grease-stained concrete around the mechanic's pit. He grimaced at Eddie. "This one's a little far gone, don't you think?"

"Not in my books." Eddie eased over to the driver's door and creaked it open. He brushed his fingers over a black identification plate affixed to the pillar post. "Especially after I found out I used to own it."

"Really? Where did you find it?"

With a glistening in his eyes, Eddie gestured with his hand out the garage door. "Aways over toward Wasaga Beach on Highway 92. She was setting behind a barn that'd just collapsed. Wasn't for that, and the leaves gone from the trees, I wouldn't have seen her."

"Wow." Angus studied the empty headlight casings. There were tarnished chrome visors above them, almost like eyebrows. Peering ahead, waiting for new glass lenses. He looked back at Eddie. "Did you know she was still around?"

"Nosir." Reaching into the pocket of his coveralls, Eddie pulled out a pair of rusted Ford keys. He jingled them on their chain. "Hadn't seen her since, oh? Since I traded her in on a new '62 Falcon." He pointed out the open door again. "Over in Elmvale. Place ain't in business anymore. Back then, I figured some local kid must've bought her. Banged her up in a figure-8 race. Then sent her to the crusher. But nope, she survived. That's rare, for a Mainline."

Angus's eyes travelled back to the '56 Ford. Eddie saw that. He stepped away from the door post and gestured toward the car with an open palm. Angus eased over to the driver's door, afraid to touch it lest the rusted panel come loose. The window glass was gone. He leaned into the front seat area with his hands clasped behind his back. The steering wheel had a dished centre. There was a console of round gauges below the dashboard. He peered back at Eddie.

—W. MacLaren Wilson photo

"I guess you're going to try fixing her up? Like she was when you had her?"

"Count on it, Angus." Eddie glanced up a wall joist. He was looking at a pair of white-on-black 1957 license plates nailed to it. Angus saw his eyes glaze over. Eddie walked around to the front of his '56 Ford and stared at the grill. "Come back in the spring. You can take Amanda for a ride."

Old Eddie's car had turned out to be one of two necessary ingredients. The other was Angus's deep-seated desire to visit the past. When the two intersected, Angus's time-travel trip became the result.

Over this past winter, Angus thought of Old Eddie many times. Always in a rather melancholy way. He figured that Eddie, at his age, had perhaps one more automobile restoration left in him. In the 1956 Ford, he picked the right car for the right reasons. But Eddie's objective seemed so far out of reach that Angus didn't have the heart to visit him for several months.

Lying awake some nights, Angus pondered what Eddie and his pre-1960 cars represented to him. It was probably their association with the 1950s. But the root of his fascination for that decade remained a mystery. A year ago, Angus attended a storytelling workshop. At the end of

—W. MacLaren Wilson photo

Tiffin yard, Midland, April 3, 1959.
—W. MacLaren Wilson photo

the first session, he was sent home with an assignment—study his family lore and come up with a tale to share with the group. Seeking inspiration, Angus browsed through cardboard boxes containing his family's photographs. They went back three generations. He selected about a dozen snapshots. It turned out they were all from the 1950s. They showed steam trains, steamships, family cars, the Riverview Inn at Washago, his grandparents' childhood homes.

With the pictures in hand, Angus returned to the storytelling group. Sitting cross-legged on the floor as part of a circle, he laid the snapshots down in front of him. As best he could, he explained his fascination for the era depicted—the 1950s. He told everyone he wanted to explore those times. But he couldn't explain why. More to the point, no family story came to mind. Sympathetic heads nodded at him, but offered no insights. Eventually, Angus told everyone a big fish story involving his grandmother. That didn't answer the question as to why the 1950s fascinated him, but it entertained the group and fulfilled his assignment.

Angus knows he isn't the only one fond of the 1950s. Nostalgia is big business. What most people refer to as the "1950s" isn't the actual decade. It's really the interval from Bill Haley & His Comets topping the charts with "Rock Around the Clock" in July 1955 through the Kennedy assassination in November 1963. Rock 'n' roll, fashion, cars, pop culture, social issues. The aspects of the decade—the actual calendar decade—that matter to Angus are different, though. For one thing, the 1950s marked the end of North America's railway steam era—that world enshrouded in coal smoke "down by the tracks".

Concerning Angus's interest in steam locomotives, people ask him a common question. *What got you into this?* He's never provided a simple response. That's because he isn't sure what the answer is. He could say he's been fascinated with steam locomotives ever since he first saw one operating. That was at a museum, when he was two or three. But thousands of kids have such experiences with no lasting effect. Angus could point to the train sets he received at

Christmas. But what child didn't received a toy train early in life? He couldn't pin it on his grandparents, either. They'd related stories of steam trains to him, in passing. But not in a way to impart a lifetime interest.

As a boy, Angus spent time around active steam locomotives. He convinced his dad to take him to see those which ran occasionally in Ontario. During family vacations to the United States, he wrangled side trips to see others. He climbed into cabs, shovelled coal, helped enginemen shake grates and blow down boilers. Felt the heat, pulled whistle cords, eased open throttles to turn six-foot drive wheels. In retrospect, the experiences have left him perplexed. Steam locomotives, while fascinating in themselves, seem more like triggers to imaginary 1950s experiences. The engines aren't what matter, per se. It's the era. That said, steam locomotives are the core of his interest in the time period. Without them, the past holds no appeal.

There's a funny thing about Angus's fascination with Eddie's old cars—they represent the downfall of the railway era. Over the 1950s and beyond, the family car produced urban renewal. North American cities and towns lost their inner cores. They sprawled into bland, vast, flat suburbs—pavement and concrete everywhere. Highway bypasses undid ties between communities. Remaining streetcar and interurban lines vanished. A driving culture prevailed. The physical world became defined by the highway. People isolated themselves from each other in their vehicles. Most towns and cities lost their railway tracks, and certainly their passenger train service. Just for a few years, though, railways and highways coexisted in a pleasing combination. That era—from the end of the Second World War through the 1950s—interests Angus the most.

Steam locomotives were Angus's entry into the 1950s. With them in place, his fascination for the era telescoped outward. He's become a student of railway operations of the period. That devotion extends to other transportation modes and industries. Cityscapes, advertising. The whole scene, really. He's also developed an interest in the working man of the era. The one who made the world operate—railwaymen, factory workers, lumberjacks, stevedores, truck drivers.

Welland Canal, spring 1959.

—W. MacLaren Wilson photo

Penetang,
July 18, 1958.
—John Rehor photo

The common man doing common tasks. Angus wants to introduce his Millennial generation to them. If he can explore the past firsthand, he can relate his findings to people of the 21st century. Like a witness to the past, a voice for unsung heroes who did everyday things.

The bigger picture of societal values means something to Angus as well. Decorum seems to have been lost, along with steam locomotives, as the world moved on from the 1950s. The family unit took a severe hit. Manners and etiquette went out of fashion. Granted, *Leave it to Beaver* offers a restricted worldview. Angus wouldn't try to argue that the times were quaint and innocent and simple. They were and they weren't. But the 1950s is the time period where he lives in his imagination. Sometimes, he wonders if his obsession with the past is a way to escape the present. To avoid reality, even. But he doesn't want to live in the past. He only yearns to visit it. The quality of yesteryear that makes it safe, despite its tragedies, is that it's already happened. By virtue of hindsight, one can visit the past and avoid its horrors.

Since Angus hadn't known the 1950s firsthand, he can only experience the era through some form of story. There's something potent about a combination of words and pictures. Sometimes, he attributes his interest in 1950s railways to a book his grandfather had. He bought it a couple of years before he died. It's called *Steam at Allandale*. It belongs to Angus now. As a boy, he read it over and over on the verandah of his grandmother's house. Between its covers, he travelled back to 1954 on an idyllic summer day. He lived in the descriptive prose and pored over images depicting steam locomotives on branchlines of Ontario. He's since bought all the books in the series and done the same with them. Through their pages, he's developed a yearning to see firsthand the twilight of the railway steam era in Ontario.

There was a clay animation television show in the 1950s and 1960s—*Gumby and Pokey*. Angus watched a number of episodes on YouTube in his boyhood. The characters in the show could step into any book they wanted. They'd physically pass through the cover on the screen. After that, they'd be living in whatever world the book described. It could be about knights and castles, Robin Hood, ancient Rome, the Old West—anything. Angus wants to do that too. The book he'd choose? *Steam at Allandale* or one of its sister publications.

Unlike Gumby and Pokey, Angus can't enter the world of a book of his choosing. But he can explore the settings they depict firsthand. He's done that. Many times, he's driven to railway towns portrayed in 1950s photographs. But, after living in the pictures in his imagination, the real-life contact has been heartbreaking. The tracks have been gone in most cases. Elapsed decades have made the scenes almost unrecognizable. That's the thing about time. It keeps moving on. Nothing stays the same. Every passing year pushes the 1950s further from living memory. One can't visit the past by travelling to a locale featured in an old photograph. It's an exercise in insanity to believe otherwise. You may as well try unscrambling an omelette.

Places can be recreated as they were in the 1950s. Museums and movie sets feature that kind of thing. One way or another, any scene can be represented. It might require digital effects such as CGI, but even the vast D-Day landings can be re-enacted. Smart technicians and artists do that stuff. But, recreating aspects of the past doesn't make it the past. Any such attempts are creations of the present. The only way to experience yesteryear would be to visit it firsthand—through time-travel. But Angus always considered that impossible. To be honest, he never considered it. The past being out of reach makes it all the more appealing. One could travel anywhere in terms of distance. But no one can travel across time.

Amanda makes Angus squirm over his contradictions about the 1950s. They both laugh about that. The era looks interesting to him until he ponders going to a bank between 9 a.m. and 3 p.m. on a weekday to get cash. Or, travelling hours to meet other people who share his interests, perhaps at monthly meetings, instead of connecting via the Internet. His grandmother would overhear him talking about the romance of old tourist cabins. They'd been located along highways in the 1920s through the 1960s. She'd say, "They were fine, unless you had to sleep in one." Ditto for Angus's zeal for steam-powered branchline trains with wooden passenger cars. "They were fine until you had to ride one." Then Amanda would get back into

Riverview Cabins, Highway 11 north of Orillia.

—Gord Launchbury collection

CPR 1004 and CNR 5563 at Harriston Junction, July 23, 1957.
—Don Wood photo

the act. Would Angus choose canned fruits and vegetables of the 1950s over today's fresh produce? Did he prefer open windows, fans and dust to air conditioning? What about postal mail to email? Waiting a week for photographic film processing versus digital images? It would go on and on. At the end of the day, though, Angus still longs to experience the bygone era.

One day, Angus tried an experiment. He wanted to see how the average person would respond to the question that he always fielded concerning his interest in the 1950s. A father of one of his baseball teammates is a heavy-equipment mechanic. So, Angus put the question to him, concerning his career choice. *What got you into this?* Just the way everyone bothers him about steam locomotives.

The dad answered that it was a job! Pressed further, he said he was good at fixing things, working with his hands. He'd started on tractors during a co-op placement in high school. Along the way, he'd learned there was good money and demand for heavy equipment technicians. The local college offered a course. It seemed like a fit, so he went in that direction. Of course, Angus's teammate gave the real reason for his dad's career direction, in confidence. Working on graders and excavators was a macho kind of job, a field for real men. It went with the old man's personality, making him feel important and tough. After Angus received those earthy answers to the question he'd long suffered, he gave up on giving people complex explanations. He stuck with the simple.

It wasn't long ago that Angus dropped in to see Amanda at her parents' home. She's slender, with long chestnut brown hair and amber eyes. A bit more than seven months older than he is.

She's a perfect combination of good looks and intelligence. Imaginative, too. She was sitting at the Webbs' kitchen table, blowing on the fresh purple nail polish on her fingers.

Taking a seat across from Amanda, Angus laid his copy of *Steam Over Palmerston* on the table in front of her. The more Angus perused the hardcover book, like others in its series, the tougher it was to make it go away. No matter what occupied him day-to-day, his thoughts kept wandering to the world depicted in the books. He opened the volume to page 129. That was something he'd done at least a dozen times with Amanda. The mid-morning sun glanced off the black & white photos on the glossy paper.

"Don't you wish you could go there?"

"Where?" Amanda looked across the table while her toes curled around the rungs of her chair. In a bit of a strangling manner, it appeared to Angus.

He tapped a picture of a Canadian Pacific steam locomotive. It had been taken at Harriston Junction in July 1957.

"What is it about these books, Wolfie?" Amanda glanced down. "Or this photograph, even?" She swept her fingers away in a backhand manner.

"The setting, the caption. It draws me in, ignites my imagination. I long to live in the tranquility of that photograph. Imagine, Amanda, two branchline steam trains meeting at a crossing. Not far from where we live. What were the chances of that happening? I want to drop everything and go where the picture was taken."

Amanda sighed. "You've dragged me to every one of the places in these pictures. And all the other books. Sometimes twice. I see them in my sleep."

"No, it's not that. Not like those times." Angus cocked his head as the whistle of a freight train sounded through the open crack of a window behind him. "I want to go to these places, well, in the summer."

"Duh." Amanda ran her hand through her hair. The strands fell midway down her back. "We've done that. Every season. Wintertime in a blizzard. Fall when the leaves were just right." She counted off with her fingers. "Spring when it's easy to see the old railway embankments. And summer, when—"

"Not just any summer."

Summer 1980.
—author photo

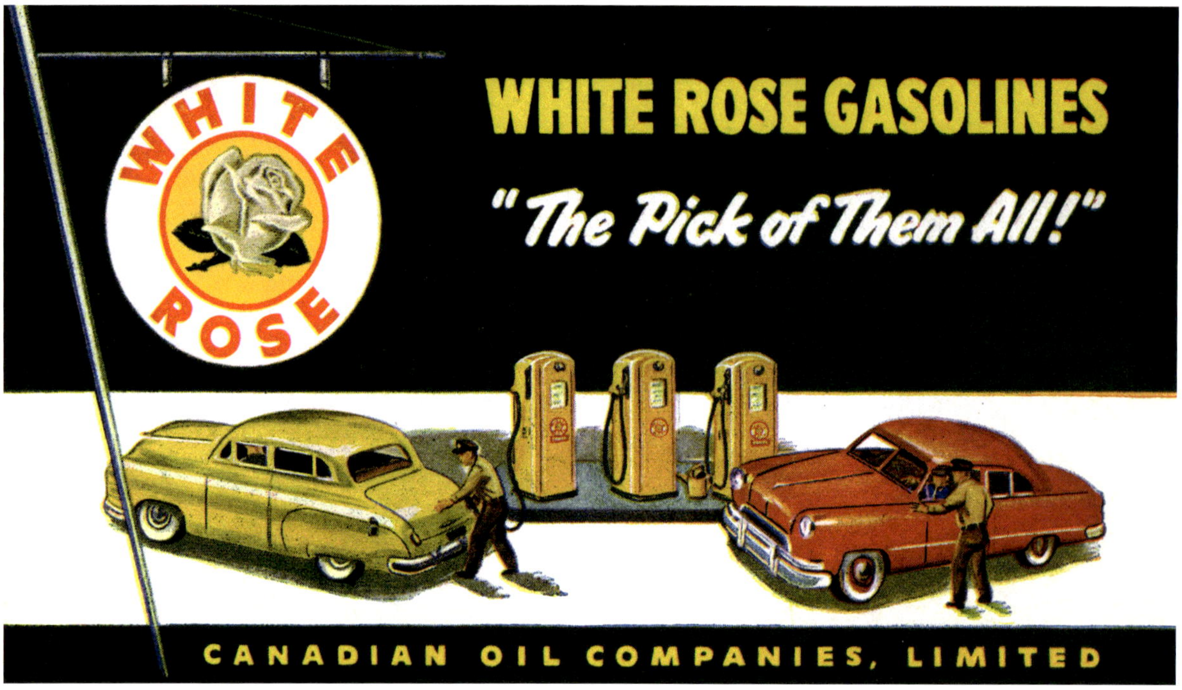

Amanda rolled her eyes. "We've gone on those outings, Angus." She only called him by his proper name when he annoyed her. "Over and over and—ugh."

"*This* summer." Angus tapped the magazine.

"You're always searching for something you can't find." Amanda clenched her fists and shook them. "Don't you get that?"

"Yes, I know." Angus exhaled and stared down at the picture in the book. "I want it to be 1957."

"It never is, Wolfie." Amanda clasped his fingers in hers. Her voice dropped. "Never. And never will be, no matter how many times we try to reconstruct these old scenes."

Pursing her lips, Amanda closed the book and handed it to Angus. "You're 21 years old. Socially-confident. You enjoy life with me. We're attached. You have other people in your life too. And you're strong and healthy with a perfect physique. You can have everything you want in the present. What's wrong?"

"I feel like I'm not fitting in. I'm not a social animal. In moments of melancholy, I dwell on this image of Harriston Junction in July 1957. I want to step into that picture. I don't know, maybe to escape the trials and tribulations of being a young adult in this day and age."

"So you want escapism." Walking around behind Angus's chair, Amanda tugged him to his feet. "Shoo, Wolfie." Guiding him from behind as if he were legally blind, she navigated him to the front door and out onto the porch. "Here's your coat. Come back when you can talk sense."

Angus would come back, later. He would indeed. After visiting Old Eddie, that is. But what he'd say then to Amanda would make even less sense.

The story developed mid-morning, a few days after Amanda had shooed Angus from her house. It was late March. Old Eddie was where he always was, it seemed—in the service bay of his former White Rose gasoline station. When Angus peered in through the open garage door, he hadn't seen his old friend in months.

"Hi, Eddie."

"What's up, young fellow?" Eddie wiped his hands on a shop rag. "I was going to call you over anyway."

"Oh?"

In his navy blue coveralls, Eddie leaned back against the front hood of a vehicle. He cocked his head and narrowed his gaze. His lips formed into a smile as he probed Angus's face.

Then Angus identified the car. He gasped.

Eddie chortled. "It's her all right."

"Wow. You weren't kidding." Over the winter, the old mechanic had accomplished the impossible. The 1956 Ford was no longer a derelict. It now sported a fresh coat of Bermuda Blue. Through the garage door, sunlight reflected off the front bumper. Angus approached the car, marvelling at its glistening chrome and restored panels. The hood ornament, badges, hubcaps—all evoked a new model. Fresh out of the factory. Around the rear, the taillights were eye-catching—round lenses with protruding centre rings.

"Angus, it feels satisfying to have restored this old girl." Eddie leaned against the garage door frame, tapping a wrench in his hands. He hesitated, prompting Angus to look over at him.

"But?"

"But she needs a proper setting in which to drive." Eddie laid the wrench on top of a gizmo with dials and wires that rested on a cart. "Cars from the 1950s in today's world are like fish out of water." He stared out the garage door at the stream of cars crossing the canal on the four-lane highway. Then he turned back to Angus. "Are you still fixated on those old trains?"

"Yeah." Angus studied the reflection of the sun's rays in the car's blue finish. "You know how once you get something into your head, you can't shake it?"

—W. MacLaren Wilson photo

"Yessir." Eddie wedged a cigarette into his mouth. He clicked a lighter cupped in his hands, igniting the tip. It glowed orange as he sucked back. He exhaled two streams of smoke out his nostrils toward the ceiling. "That's the only way I can fix something like this old girl." He tapped the left front fender of the Ford behind its visored headlamp. "By being bull-headed about just the one thing."

Angus smiled. "That works when you're restoring an old car." He touched the hood and pulled his fingers away. Even though the metal was shaded, it was hot. Like it had been sitting out in summer glare. "Why didn't you tell me this thing was cooking?"

"Hah." Eddie cackled and took another drag on his cigarette. "So, you were going to say something to me. About getting hung up. What was it?"

Angus scrunched up his face. "How do you go about achieving the impossible?"

"Depends what the impossible is." Eddie angled his head toward Angus. "Girl trouble?"

"Not really." Angus chuckled. "But sort of. Amanda did kick me out of her house a few days back. Thought I was going batty."

"Are you?"

"Probably."

They both laughed.

Eddie stubbed out his cigarette in a rusted coffee can filled with sand. "Join the club. I thought I was cuckoo myself. Maybe I am. You can be the judge of that." He gestured with his forehead toward the '56 Ford.

"Well, you weren't." Angus squatted down and stared at his image. It appeared distorted in the car's rocket-shaped hood ornament. "Fixing up this old girl was a perfectly sane thing to do. What I want to do is just plain wacky."

—author's collection

"Try me." Eddie's voice rasped. "How wacky?" He picked up a smudged Canadian Oil Companies desk calendar that lay on a stack of old tires. Its motif was the White Rose boy-with-slate. "Tell me how wacky and I'll top it."

Angus guffawed. "Forget it, Eddie. Even you're not that crazy. I told Amanda I wanted to see a bunch of old trains. She figured what I really wanted was to go back to 1957. I guess she's right." He scuffed his shoes on the floor. Decades of oil stains and sweeping compound crusted its surface. "You can't top that one."

Eddie tapped the calendar in his hands. "Yes I can."

"No." Angus shook his head. "Nothing's more far-fetched than going back to 1957."

"Except." Eddie paused and didn't speak again until Angus's eyes met his. "Except a car that will take you there."

Angus snickered. "What, in some sort of re-enactment? You're involved in another 1950s movie shoot?"

"Nope." Eddie nodded toward his '56 Ford. "This old baby has turned out to be..." He swivelled, then walked toward the device with all the dials and wires. It rested on a cart near the hood of the car. "Know what this is?"

Angus eyed the contraption. "Something from Dr. Frankenstein's lab? I don't know, Eddie. What is it?"

"Just a basic motor analyzer." Eddie pointed to its array of instrument panels. "You hook it up to an engine. Measure all kinds of things. Compression, pressure, amperage, ignition, timing. Every shop had one back in the 1950s."

"Okay." Angus scrutinized a gauge that assessed the degree of engine vacuum, in inches. "And you used this one to tune up your old car?"

"Sort of." Eddie moved beside Angus. He pointed a smudged finger to a spot on the calendar he was holding. "See this date?"

Angus nodded. "Yes. July 1957. From when your dad ran the garage?"

"No. But that's not the important thing." Eddie ran his finger along a week's worth of days starting with the 21st of the month, a Sunday. "See all them letters and numbers?"

"Yes." Angus frowned at the scrawled entries under various dates. None of them made any sense to him. "Some kind of code?"

"Kinda." Eddie set the calendar on the cart where the motor-testing apparatus rested. "Mechanics' code, you might say. The figures tie in to readings on a gizmo like this."

Angus shrugged his shoulders. "I'm not sure where you're going with this, Eddie. They're just some data that a mechanic jotted down in July 1957." When he heard himself say the date out loud, he felt a tingling at the back of his neck. "To tell him what was wrong with the engine, right?"

"Ordinarily, yes." Eddie unhooked a pair of Ford keys from a nail in a wall stud. "But just for the hell of it, I adjusted everything on this inline six cylinder. Until all the readings matched the ones on a day from the calendar. Just to see what our mechanic friend was listening to."

"But that would depend on the car, wouldn't it?" Angus nodded at the calendar. "It must have sounded different for every car. You don't know which one he was working on."

Eddie grinned, showing his yellowed teeth. "Except that this calendar—and others—was in the trunk of the car when I reacquired it. And"—he tapped his fingers on a string of characters penned near the top of the calendar—"this happens to match the serial number."

"Hmm." Angus cleared his throat. "So the owner of the car kept technical records on it."

"Except I owned the car in July of 1957. And this isn't my calendar." Eddie paused to let his words sink in. "Whoever wrote this stuff did it after I traded her in. That was fall 1962, almost Christmas."

"Okay." Angus swallowed. "So, you matched the readings on the calendar with the engine. That must have made it run rough. Like the other guy would have heard."

"Nope. That ain't what happened at all." Eddie cackled. It was a drawn-out release of tension. "She didn't run rough. Far from it." He heaved with laughter so hard that he doubled over at the knees. When he came up, his eyes were moist. "Like I said, Angus, I can top you. This car"—he reached for the driver's door handle and pressed his thumb on the latch button—"took me back to July 1957."

—W. MacLaren Wilson photo

—author photo

"Whoa." Angus inched closer to his companion and spoke in a near-whisper. "What do you mean, Eddie?"

"Just what I said." Eddie opened the car door, which moved silently on oiled hinges. He sat down sideways behind the steering wheel, resting his feet on the concrete floor. "I matched the engine to all the readouts the guy had written down. That's all set now. It's locked in unless she goes out of whack. After that, I nosed the idle up and down a bit. Just to see how she sounded, you know?"

Angus nodded.

"And then the damnedest thing happened. I was instantly"—Eddie snapped his fingers—"back in the past. It turned out to be a summer day in July 1957. The one on the goddamned calendar."

Angus angled his head, looking away. He took a deep breath. Staring at the crumbling pavement on the abandoned section of highway, he tried to process everything Eddie had said. Then it hit him. He turned back to the old mechanic.

"You've been talking to Amanda, haven't you?" Angus growled. "You had me there for a minute. It's the kind of thing she'd dream up and get you in on. To get me to stop yakking about those steam train books."

Eddie heaved himself up from the driver's seat of the car. He came up beside Angus and rested a hand on his shoulder. "See for yourself, then. Climb in."

Angus looked at the car, then at Eddie. The old mechanic's face betrayed no hint of a suppressed joke.

"Sit down behind the wheel." Eddie handed the keys to Angus. He'd burnished them to a shine. "Close the door and start the engine."

Shrugging, Angus sat down on the new grey upholstery behind the steering wheel.

"She ain't feisty, like that little Mazda 5 you take your Amanda around in." Eddie grasped the door handle. "Only 137 horsepower, inline six engine. Rear-wheel drive. That might take some getting used to."

Angus chuckled. "You're talking as if I'm the one who's going to be driving her."

"And don't roll the window down, whatever you do. It'll kill the process." Eddie closed the door. It made a satisfying clunk of 1950s Bethlehem steel.

—author's collection

Settling back in the seat, Angus took stock of the '56 Ford. The steering wheel was huge. Of course it was. It had to provide torque without power-assist. There was little in the way of plastic in the interior. His nose caught a smell, characteristic of old cars, which defied description. It was a pleasing blend of aged materials.

"Insert the square key in the ignition, to the left of the steering column." Eddie yelled through the side window glass. Angus followed his instruction. "Now give her a little choke"—Eddie pointed to a knob beside the ignition—"and start her up."

Angus found that depressing the clutch, unlike the one on his Mazda, took some leg strength. He turned the square key in the ignition, and gave the '56 Ford a little gas. The pedal was twice the size of his foot. With a fluttering sound, the engine caught. The car shuddered, as if a Jerry Lee Lewis song had suddenly possessed it. *You shake my nerves and you rattle my brain*. It was alive.

Eddie's voice boomed above the sound of the motor. "Lay off the choke now."

Angus pushed the knob in.

"A little more accelerator." Eddie rotated his wrist, as if opening a tap.

Gradually, Angus depressed his foot on the gas pedal. The engine rpms climbed with the needle on the tachometer. He felt the throb through his hiking shoe.

Eddie clutched his fingers in a fist.

The glass all around Angus began to frost over. It was an unnatural sensation. What used to be transparent was now approaching opaque. With visibility diminishing, he saw Eddie twist his arm like a baseball umpire signalling an out. A booming sound came from somewhere. The engine? Outside? Angus shut his eyes. Hyperventilating, he released his foot from the accelerator. The car's shaking settled down to a mild vibration. Elvis Presley now. *Love me tender*. That was a relief. He opened his eyes, looking for Eddie.

But Eddie wasn't there. Nor was his garage. Or the last snows of winter. Angus gaped out the windshield, which was transparent again. Blue summer skies were everywhere. Where was he?

Kitchener, Ontario. That's where Angus was. He was parked in a gravel lot. Out the passenger window, he saw a sign hanging on the Canadian National station. In front of him, steam engine 7411 was pulling some boxcars. Wigwags at a railway crossing started swinging. The locomotive went over the roadway, bell ringing. Angus watched the engine switch some cars into a siding at a freight shed. Then it came back across the roadway and stopped.

The five-man crew gathered on the ballast near the locomotive. It looked as if they were taking a break. Angus grabbed the door crank and climbed out of the car. He looked up and down the tracks, saw no trains coming, and approached the train crew. He asked the men about their engine. They said she was almost out of service, although her tender had been repaired at Stratford recently. Indeed, it had, Angus could see. There was a fresh red and yellow herald emblazoned on it. The engineer said his locomotive would probably be equipped with a steam pipe. Then she'd round out her days melting ice from refrigerator cars at the Stratford yard. After that? The scrapper's guillotine at London.

Angus noticed a switchlist in the conductor's pocket. He tilted his head and read the date scrawled across the top. July 6, 1957. Just as Eddie had said.

The encounter at Kitchener was so brief for Angus that it felt like a daydream. He conversed with the train crew, after which they returned to work. Angus climbed into the Ford. Then he and the car were back in the service bay of Eddie's garage. He cranked the window down. Eddie leaned in, grinning. Angus got out of the car, teetering on his feet. He heard his own voice

Kitchener, July 6, 1957.
—John Rehor photo

Kitchener, July 6, 1957.
—John Rehor photo

babbling about a steam engine and coal smoke at Kitchener. No amount of drunkenness could have made him that incoherent. And, to Eddie, it appeared as if he hadn't gone anywhere. What had been a time-travel experience for Angus had been but an instant to the outside observer.

Needless to say, Angus peppered Eddie with questions after the Kitchener experience. Standing in the driveway of his White Rose station, the mechanic swore that when he owned his '56 Ford the first time—over the second half of the 1950s through most of 1962—it had no special qualities. Nor had he. The car ended up on a farm west of Elmvale. Sometime in the 1970s, the province expropriated the land to accommodate a highway widening yet to occur. The owner vacated the property but the buildings and car remained. Upon discovering the car the previous autumn, Eddie purchased it at scrap value.

Eddie had moved the derelict to his garage back in the fall. Angus had seen that. Over a number of months, Eddie restored his '56 Ford. Just as he had many others of its vintage. He reconditioned some parts and added replacements as needed. Along the way, he researched the car's VIN—Vehicle Identification Number. He discovered that someone had purchased his car in 1963, from the dealership where he'd traded it in. The last registration on the provincial ministry's books was for 1968. After that, the vehicle presumably sat abandoned in the field.

It's that almost half century of the car's laying idle which is of most interest to Angus and Eddie. When Eddie began restoring the vehicle, he found a number of calendars—like the one he showed Angus—in its trunk. They'd been protected by an old car blanket. Their years are 1955 through 1962—the original period Eddie owned the car. But they aren't his calendars.

Angus and Eddie have looked through the calendars. Pencilled on some days of the various years are sets of numbers and letters. As Eddie says, they denote engine characteristics as

—Ron Johnson photo

measured by a motor analyzer. If he matches the characteristics of his '56 Ford to the readings, then sets the idle to the right number of rpms, he and the car go back in time—to the date on the calendar. There appears to be only one location for a given date. For example, if Eddie matches the idle to the reading for July 6, 1957, he ends up where Angus did—in Kitchener. He sees locomotive 7411 switching boxcars. Eddie has only checked out a few entries, all from July 1957. When he comes back to the present, he pencils the locations and times, where he ends up, on the calendar. That information is there with the engine diagnostics.

Well into the wee hours of a couple of nights, Angus and Eddie have discussed the 1956 Ford Mainline and its powers. They've concluded that some unseen hand marked up the calendars while the car sat uncertified—but engine operable—in the farmer's field. Whose was the unseen hand? Somehow, that person discovered the car was a time travel vehicle. Or made it one. Whoever it was might not have gotten very far with his expeditions—or wanted to. The vast majority of dates on all the years are blank. Angus and Eddie are thinking the same thing about that—maybe the '56 Ford can travel back in time to places as yet undiscovered.

The mysteries surrounding the trips to yesteryear are perplexing. But the traveller has control over the triggering mechanism—the engine idling speed. Curiously, the process won't work if a window in the car is open. That suggests a pressure aspect. Angus and Eddie are still trying to deduce what makes the car and occupant return to the present. If the driver doesn't do anything adventurous, his time in the past expires at the same time every time. It's as if something runs out. Like a battery with only so much juice, or a clock wound a few turns. Unlike the trip to the past, the return to the present doesn't require the engine to be operating. Or the driver to be in the car. Eddie has driven around on a couple of occasions in July 1957. Both times, vibrations from other sources brought him back earlier than usual. Once, it was a power line humming. Another time, a truck going by. Those suggest some kind of force field at work—electrical, mechanical, magnetic, atmospheric—to propel him and the car back.

M606 at Bobcaygeon, July 6, 1957.
—John Rehor photo

Angus tried the Kitchener experience a couple more times. He tried to get a feel for his life in the past, compared to that in the present. Were they separate or the same? So far, the best analogy he's found is that of taking a vacation. Like his recent one to the Florida Keys, with Amanda, fishing and beachcombing. A short while after he returned to snowbound Canada, it felt as if he'd never gone away. There were no reference points to link the two realities. And that's just like the sensation which remains after experiencing time travel in Eddie's car.

Of course, Angus had enjoyed his brief trip to Kitchener in July 1957. A week ago, he'd dropped in at the White Rose station. "Do it again, Eddie."

"After I've given you a bit of instruction on operating the '56 Ford." Eddie laid a drop cloth over the passenger half of the front seat. "Don't want to mess up the fabric with grease from my coveralls." He tapped a buckle beside him. "I liked the '56 Fords because they were the first with seatbelts. As an option, mind you. Only twenty percent had 'em, but you're lucky. And she's got a good twelve-volt battery, another first for the line." He rapped the rounded edge of the dashboard. "This car hit the dealer's lot in September of '55."

Angus tapped his teeth together. "So that's as early as we can probably go."

"If we get the readings rigged." Eddie hacked a smoker's cough. He tapped the shift lever on the right side of the steering column. "Toward you and down for first, forward and up for second, straight down for third." He pointed along the driveway leading from the old White Rose station. "Take her out on the road there, to the left. And Angus?" He narrowed his eyes to slits. "Don't put us in the canal."

After Angus had been "checked out" on Eddie's 1956 Ford, he visited his old mechanic friend again. This time he brought Amanda along. He only told her that Eddie had restored his car and taught him to drive it—not that it had special capabilities. Angus prearranged with Eddie to make the next trip alone, with Amanda only as a bystander. Neither had any desire to freak her out too soon.

Eddie consulted a second set of entries for July 6, 1957. It called for a different idling speed, but otherwise the engine readings were the same as for the Kitchener trip.

Angus climbed into the car, started the engine, and eased the idle up. He kept his eye on Eddie. The rpms approached the sweet spot. Then—the frosting glass, the booming sound. And there he was. This time, in Bobcaygeon. A village in cottage country northeast of Toronto.

The '56 Ford was parked near the Canadian Pacific depot. Angus climbed out of the car. Over by the station was a Ten Wheeler steam locomotive, number 434. The engine was resting, with a passenger-and-baggage car, on a track in front of the car. Angus watched the station agent unloading express packages from the train. There was a big flat carton from Eaton's. Also, wooden cases of something called *Wishing Well* soda. The temperature was perfect, the skies clear, the breeze refreshing. Over on the face of the station, Angus discerned a chalkboard. He checked the date at the top. It was July 6, 1957. Sure enough, it was the same day as at Kitchener, only more toward noon.

Like that visit to Kitchener, the trip to Bobcaygeon lasted only a few minutes. Angus then found himself back in the driveway of the White Rose station. He smiled at Eddie and Amanda, who were standing next to the driver's door.

Amanda must have noted a change in Angus's face. It prompted her to voice suspicions that something had jarred him. He watched her scrutinizing him. Her face tightened. In a shrill voice, she laid into Old Eddie for messing with Angus's mind. That prompted Angus and Eddie to explain to her what had happened. But Amanda wasn't swallowing any of it. She said she was onto the pair. The whole time-travel thing was a prearranged joke they'd rigged up. It was all pretend. Angus understand her skepticism—it mirrored his own of a few days earlier. He knew there was one way to allay her suspicions. So did Eddie.

—author photo

M605 at Bobcaygeon, July 12, 1957.
—John Rehor photo

"Climb in beside Angus." Eddie opened the passenger door of the '56 Ford.

Angus studied Amanda's face as her eyes peered into his. A patronizing smile formed on her lips. She sat down in the front seat and reached for a shoulder belt that wasn't there. The same thing he'd done. Eddie nudged her door closed. From behind the steering wheel, Angus watched Eddie grab his 1957 calendar. He turned to the month of July and positioned his finger on some pencilled characters. They appeared to be six days later than before. That made it the twelfth of the month.

Once again, Angus started the engine of the '56 Ford and gradually revved it up. He looked out the windshield. Eddie, in his grease-stained coveralls, leaned an elbow on the car's hood. He cocked his head toward the occupants, gesturing with his free hand for more accelerator. Angus increased the pressure with his foot. The engine frequency increased. He looked over at Amanda as the motor rumbled. The glass began to frost over. He detected a humming sound. It gradually focused on a single pitch, like a tuning fork. The rpms were approaching the sweet spot. Angus breathed a goodbye to the March morning. Eddie clenched his fist like the baseball umpire. The bang sounded. Amanda gasped. Just as she did the first time she took the plunge on Disneyworld's Tower of Terror.

Then, all was quiet. Angus looked out the windshield onto a summer day. They were parked at the end of a dirt road by a lake. He gripped Amanda's hand. She was trembling. Then he heard a bell clanging and something chuffing. There was a steam locomotive passing them on a track to their right. It was pulling a three-car train across a bridge. The engine was Canadian Pacific's number 484. Just like the other one he'd seen at Bobcaygeon. Amanda's jaw dropped. The previous summer, in their regular lives, they'd popped into the Exporail museum, near Montreal. One display had been a magnificently-restored sister engine—number 492—coupled to a similar train. But those pieces had been static. The scene unfolding from their vantage point was not. Locomotive 484 exuded the quality of life.

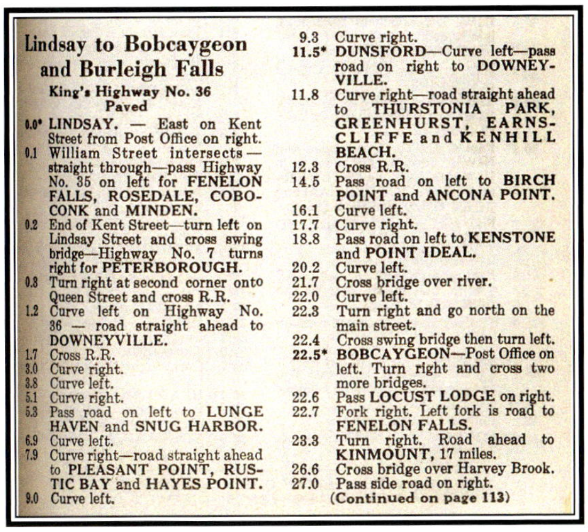

Lindsay to Bobcaygeon and Burleigh Falls

King's Highway No. 36
Paved

0.0*	**LINDSAY.** — East on Kent Street from Post Office on right.
0.1	William Street intersects—straight through—pass Highway No. 35 on left for **FENELON FALLS, ROSEDALE, COBOCONK** and **MINDEN.**
0.2	End of Kent Street—turn left on Lindsay Street and cross swing bridge—Highway No. 7 turns right for **PETERBOROUGH.**
0.3	Turn right at second corner onto Queen Street and cross R.R.
1.2	Curve left on Highway No. 36 — road straight ahead to **DOWNEYVILLE.**
1.7	Cross R.R.
3.0	Curve right.
3.8	Curve left.
5.1	Curve right.
5.3	Pass road on left to **LUNGE HAVEN** and **SNUG HARBOR.**
6.9	Curve left.
7.9	Curve right—road straight ahead to **PLEASANT POINT, RUSTIC BAY** and **HAYES POINT.**
9.0	Curve left.
9.3	Curve right.
11.5*	**DUNSFORD**—Curve left—pass road on right to **DOWNEYVILLE.**
11.8	Curve right—road straight ahead to **THURSTONIA PARK, GREENHURST, EARNSCLIFFE** and **KENHILL BEACH.**
12.3	Cross R.R.
14.5	Pass road on left to **BIRCH POINT** and **ANCONA POINT.**
16.1	Curve left.
17.7	Curve right.
18.8	Pass road on left to **KENSTONE** and **POINT IDEAL.**
20.2	Curve left.
21.7	Cross bridge over river.
22.0	Curve left.
22.3	Turn right and go north on the main street.
22.4	Cross swing bridge then turn left.
22.5*	**BOBCAYGEON**—Post Office on left. Turn right and cross two more bridges.
22.6	Pass **LOCUST LODGE** on right.
22.7	Fork right. Left fork is road to **FENELON FALLS.**
23.3	Turn right. Road ahead to **KINMOUNT,** 17 miles.
26.6	Cross bridge over Harvey Brook.
27.0	Pass side road on right.

(Continued on page 113)

Angus watched the side rods of the locomotive moving up and down. It crossed the bridge and an adjoining trestle. He and Amanda reached for the cranks and lowered their windows at the same time. A shore breeze wafted a trace of coal smoke through the front seat. The train included two freight cars—a keystone-heralded Pennsylvania boxcar and a Nickel Plate Road gondola. Bringing up the rear was a Canadian Pacific combination car, for baggage and passengers. Sunlight reflected off its wine-coloured finish. There was a man sitting in the left side of the engine cab, above the yellow number 484. He waved at the two of them. Angus's knowledge of 1950s railway crews told him he'd be the fireman or head-end brakeman. Raising his hand, Angus returned the greeting.

Amanda waved at the engineman too. She turned to Angus. Her face was losing a bit of its tension. "Where are we?"

"Near Bobcaygeon, I think." Later, Angus learned that they had been half a mile short of the village's railway station. They had been parked on a road in the cottage community of Kenstone Beach. It was on Sturgeon Lake. The train had crossed Little Bob Channel.

Angus grabbed the gearshift lever and steering wheel. He negotiated the '56 Ford around the end of the dirt lane. They headed off in the only possible direction. The visit back in time probably wouldn't last long. He wanted to make the most of it. They had a train to chase. And the great thing was—no one else cared. Not like in his normal world. There, every appearance of a steam locomotive generated a horde of camera-clicking disciples.

While tearing after the train, Angus grasped a principle about entering the 1957 scenes. The sudden appearance of the '56 Ford and occupants never seemed to jolt anyone. Nobody in the past gave the impression that they had taken note. It was as if the car and passengers were there all along. Was it the same when they disappeared? Maybe their comings and goings were like the proverbial tree falling in the forest. Not making a sound if no one was there to witness it.

At the end of the trail, Angus made a left turn onto a concession road. That took them over the track and then to King's Highway 36. With an eye on a smudge of coal smoke, he turned left. Good thing they were chasing a Ten Wheeler on a branchline. According to Eddie, the top speed of the '56 Ford was 91 m.p.h. That was slower than CPR Pacific Hudson steam locomotives pulled passenger trains. Angus took note of that thought. He was already considering 1957 to be the present.

M606 at Bobcaygeon, July 12, 1957.
—John Rehor photo

Two miles of open-country driving brought Angus and Amanda to the village of Bobcaygeon. He recognized the spot from his recent 1957 visit. Funny, the two of them had also driven through there the previous June. That had been in their regular lives. They'd strolled by the Trent Severn locks. Also, down along the waterfront on Park Street, where they were now driving. Angus manoeuvred the '56 Ford into a gravel parking lot and turned off the engine. The railway station was to their right. It had certainly not been present in their modern-day visit to Bobcaygeon. The red finish on the structure matched that of the CPR passenger car.

Angus and Amanda climbed out of the car and strolled over to the railway station. Two people, who looked as if they'd just gotten off the train, were walking along the platform. There was a man who appeared to be a postmaster. He was taking bags of mail from a crewman in the baggage compartment of the passenger car. Another fellow, who looked like an express clerk, stood nearby with a wagon. He had several cardboard packages on the platform which looked ready to be loaded aboard the train. Someone inside the baggage compartment was handing him cases, which he was stacking on the wagon. Further down the platform, behind the locomotive, the boxcar door was open. Angus could see packages of freight resting on the floor of the car.

As Angus and Amanda walked along the wooden station platform, they heard a clicking sound. He pointed through the facing window of the operator's bay. "It's the telegraph key." They stopped and listened as the man inside tapped out a rhythm. On the desk was a handwritten note. Angus figured the operator must have attached it to the waiting room door awhile back. It was in a flowing script. At cottage. Back at train time.

TORONTO—LINDSAY—BOBCAYGEON

Via Dranoel

36-M605	Miles	TABLE 40	35 M606
A.M.		(Eastern Time)	P.M.
† 9.20	0.0	Lv.....Toronto‖ 38.....Ar	† 7.30
† 11.05	62.1	Ar........Dranoel........Lv	..
..	..	Ar.......Peterboro.......Lv	† 5.15
† 11.05	0.0	Lv........Dranoel 38....Ar	..
ƒ 11.11	2.2	..........Bethany.........	† 3.56
ƒ 11.21	6.3	.........Fleetwood..........	ƒ 3.46
ƒ 11.34	10.7	............Rea's............	ƒ 3.34
ƒ 11.40	13.0	.........Hillhead.........	ƒ 3.28
12.01	18.1	Ar........LINDSAY........Lv	3.15
12.25	18.1	Lv........LINDSAY........Ar	2.42
ƒ 12.37	22.4	......Pleasant Point......	ƒ 2.31
12.48	26.4	.........Dunsford..........	2.21
ƒ 12.58	30.3	........Ancona Point..........	ƒ 2.11
ƒ 1.02	31.6	........Birch Point........	ƒ 2.07
ƒ 1.06	33.1	.........Kenstone.........	ƒ 2.03
† 1.15	34.3	Ar.......BOBCAYGEON.......Lv	† 2.00
P.M.			P.M

Someone had left a metal toy jeep on the adjoining bay windowsill. "It's wet." Amanda brushed a water droplet off the side of the toy. "Probably belongs to a little boy who was playing at the beach."

Angus fiddled with the jeep's steering wheel. "Hey, the front wheels actually turn."

Just then, the waiting room door of the station opened. Angus and Amanda gave a start. Two men walked out. By the way they were dressed, one was the locomotive engineer and the other was the conductor.

"They probably just signed the train arrival register." Angus watched the men walk back to their train. "Let's have a look at that." He pointed out a schedule board to Amanda. It hung on the trackside face of the station. They moved close enough to read the chalked inscriptions, including the date. July 12, 1957. According to the information, the train that had arrived at 1:15 p.m. Eastern Standard Time was M605. It would leave for Lindsay 45 minutes later as M606. Angus reasoned the "M" must stand for "mixed", as in freight and passengers combined.

Amanda furrowed her brow. She was looking all around her. Angus could see her coming to the realization that they were back in time. Maybe it was her familiarity with the village of Bobcaygeon that convinced her. A rational voice inside her head had probably been holding her back. Perhaps reasoning that they'd stumbled upon a historical re-enactment or rehearsal for a movie. It was funny what people notice in such moments of awakening.

"Look." Amanda pointed to clothes which hung on a line stretched between a guy wire and a pole. "The station agent's wife must have hung them."

"Yeah." Angus watched the clothes fluttering in the offshore wind. "They live in the station, by the looks of things. Good thing the breeze is taking the coal smoke away from the laundry." A thought hit him. "So many people our age want to travel. You know?"

"Tell me about it." Amanda ran her fingers along the wood facing of the railway station. It appeared to Angus as if she was still trying to ground herself in physical reality. She spoke in a quiet voice. "I made a bucket list of places once."

"Oh?"

"Everybody does, Wolfie." Amanda looked over, flicking her hair out of her eyes. "Climb the Himalayas. Swim with dolphins. Camp out with gorillas. Soak up the Greenland tundra with aboriginals."

—Hookah Hopkin photo

Angus nodded at the train. He watched one of the crewmen uncouple the locomotive and signal the engineer to pull ahead. "But we can see things like this, with Eddie's time machine. Stay in our own backyard, so to speak. Explore the fourth dimension, rather than run all over the world."

Amanda laughed. "Take that, all you millennials."

"I'm serious." Angus led Amanda back to the car. They leaned on the hood, soaking up the summer sunshine and breeze. "Journeying back to 1957 beats any exotic thing you could imagine."

To the right of Angus and Amanda, the Ten Wheeler was now on the siding. It was the second track from the station. The engine backed past the onlookers. There were four men in the cab. The engineer, hand on the throttle and looking back along the track, waved at Angus and Amanda. They watched the locomotive go off into the distance and around a bend. Then they heard a putt-putt sound. A railway motor car had stopped on the siding in front of them.

"Want to ride the jigger?" The section man aboard the motor car yelled above the sound of the engine. He waved to Angus and Amanda, beckoning them aboard.

"Let's go." Taking Amanda by the hand, Angus led her to the motor car. They climbed on.

"Not from 'Caygeon, are you?" The section man looked at Angus and Amanda, who shook their heads. He put the contraption in gear and they rattled up the track in pursuit of the locomotive.

Jostling along the siding, the motor car bent down strands of grass growing between the rails. Angus studied their surroundings, comparing them to the modern day. When they'd explored the vicinity in their regular lives, Amanda had remarked upon a concrete ring. It had been some 25 yards in diameter, embedded in the ground. Around it was a mobile home park. Curious about the circular foundation, Angus had later checked in with Hookah Hopkin. He was a buddy who lived near Washago. Hookah was a reclusive Harley driver of some 70 years. He restored old pop machines and the like. Also, he was a repository of every imaginable artifact and scrap of knowledge from the steam era. Hookah had told Angus that the mysterious concrete ring was the foundation of the old railway turntable pit.

Bobcaygeon, September 4, 1958.

—James A. Brown photo

And now, there was the turntable itself. It was intact. There were no mobile homes in sight, though. They were decades in the future. As the jigger came to a stop, locomotive 484 backed onto the turntable. Three of the men climbed down from the cab. Angus and Amanda stepped down to the cinder ballast and thanked the section man for the ride.

The train crewmen began shoving against wooden handles at opposite ends of the turntable. By sheer muscle power, they were spinning the locomotive around 180 degrees. The scene reminded Angus of a picture he'd seen of a sister engine in the same spot. "Look, another one of my engines." He pointed to number 484's builder plate as it caught the glint of the sun. It was bolted to the side of the boiler below the smokebox. Class D-4-g. Built by the Angus Shops in 1914. The matter was an ongoing joke with Amanda.

One of the crew locked the turntable in place, then signalled the engineer. Locomotive 484 backed off the turntable and stopped. Angus watched the railwaymen grab their lunch pails from the cab. They walked off toward the town beach along the nearby shoreline. It was time to eat. What would those fellows be devouring in 1957?

Angus and Amanda walked a dirt road—more like a path—in the direction of the car. Behind them, the CPR Ten Wheeler ran out on the mainline. A crewman turned the switch, then the engine backed past the pair toward the station.

For the next while, Angus and Amanda watched the locomotive and crew switch cars among various tracks. The moves were accompanied by hand signals, crashing coupler knuckles, and whistling and chuffing from the locomotive. The engine pulled the Pennsylvania Railroad boxcar from the inbound train. It was taken to the freight shed track and exchanged there for a Canadian Pacific counterpart. That car had a merchandise label stapled to a placard on its side. The train crew switched it into the outbound train, first behind the locomotive.

M606 at Bobcaygeon, July 12, 1957.
—John Rehor photo

Angus heard the station operator refer to the CPR boxcar as the "way car". The empty Nickel Plate gondola was moved ahead of the passenger car in the train. The mixed train must have picked it up from a station on the way in. Maybe it had brought in a load of gravel for a roads department along the branchline.

Then, all went quiet at the Bobcaygeon railway yard. The railwaymen went to huddle in the passenger car. Angus could see them through the open end door. Talking, smoking, casting the occasional glance toward him and Amanda. That almighty regulator on the railway—Time—presided over the scene. Otherwise? Nothing. Just a lazy afternoon in July 1957. For a moment, Angus had a hankering to cross the tracks with Amanda to the town beach. They could swim and play in the sand. Soak up the tranquillity that contrasted with the noise and bustle of 21st century resort towns.

"Would you like to ride the train down the line?" The station operator was standing beside Angus, looking at him. "You won't have many more chances. Passenger service is ending this fall." He nodded at Amanda. "Maybe get your girlfriend to drive down Highway 36 and meet you at Dunsford?"

"That's very kind of you, sir."

"Call me Bob."

"Thanks, Bob." Angus bit his lip. Under no circumstances did he want to separate from Amanda in 1957. Maybe this was even a trap, however cloaked in friendliness. He tapped his wrist, where a watch would rest. "The thing is, we've got to get back to Washago." He imagined the Imperial Oil roadmap in his mind. "So we've got to drive through Fenelon Falls."

"Maybe some other time then." The man bowed his head slightly and walked back toward the operator's bay of the station.

Concerning time, Angus pondered that the Ford was giving them a lot in 1957. Certainly more than his previous visit to 'Caygeon. That was a bit unnerving. He heard the conductor's voice ring out an "All aboard" from the station platform. Standing on the gravel surface, he and Amanda looked up the tracks. The headlight of engine 484 came on. Mixed train 606 was ready to leave. Puffs of steam rose from the locomotive's whistle. A second later, two melodic wails sounded. Angus clutched Amanda's hand. Clouds of smoke billowed from the engine's stack.

Canadian Pacific's 484, a 1914 product of the railway's Angus Shops, was starting up. Barely a year from the scrapper's torch, the Ten Wheeler approached the bystanders. The locomotive swivelled from side to side with every revolution of her drive wheels. What a splendid sight, Angus thought. The headlight, ringing bell and heat waves rising from the steel rails mesmerized him. He drew an arm around Amanda's waist. With the fingers of his other hand, he pinched himself. Was the moment real? Could life ever be that good? Now, CPR 484 was gathering speed. The fireman was waving a greeting—or a farewell. Two friendly blasts sounded from the whistle as the locomotive passed. Now she was going away, heading down the line for Lindsay, Peterboro, Havelock. Pulling her three cars along a grass-grown right-of-way that would become a mobile home park. The people of the future could have that. Angus preferred the chuffing engine and the aromatic smoke that was blocking the sun's rays. Indeed, the whole uncluttered world of blue skies, green trees and red railway cars. And a steam locomotive. That was Bobcaygeon in that moment on July 12, 1957.

Angus watched the rear car and trail of smoke from the train recede into the distance. Finally, it disappeared from his sight. From time to time, the whistle carried along the shore of Sturgeon Lake. He thought about what Hookah Hopkin had told him about the declining years of the branchline. The station operator was right. Before that very October was out, the daily-except-Sunday mixed train would be a thing of the past. The little steam engines would serve Bobcaygeon as required in freight service until the winter of 1959. By mid-1961, the township would be ready to tear up the tracks. Now a rumbling sound disturbed the quiet, over near the freight shed. Angus looked. How fitting. A Smith Transport delivery truck was pulling

M606 leaving Bobcaygeon, July 12, 1957.
—John Rehor photo

up to the building. He glanced at Amanda, then the two walked back to the car and climbed in.

"What now?" Amanda looked at Angus. "Are we stuck here? This is too weird, Wolfie."

"Guess we can at least chase the train." Angus started the car. He moved his foot up and down on the accelerator, listening to the changing pitch. Looking over his shoulder, he shifted into reverse to back out of the station parking lot. Then he stopped. "Do you hear that?"

Amanda cocked her head. "Yes. Some kind of humming?"

"From over there." Angus pointed toward the trees along the waterfront. "Heat bugs, I think."

The noise from the bugs became louder. It sharpened into a whine. Angus saw Amanda's lips moving, but he couldn't hear her words for the blare of the insects. By the way her chest rose and fell, she was hyperventilating. She grabbed at the window crank and pulled it as if it were a door lever. Angus couldn't blame her. They looked similar. He reached over to calm her. At that moment, everything went quiet. The silence itself was deafening. They were back in Eddie's driveway. It was March again.

Looking around at the grey skies of early spring, Angus felt a delicious sensation sweep over him. Eddie's time-travelling car had worked again. The phenomenon was repeatable.

That second visit to Bobcaygeon prompted Angus to explore advanced issues of time travel. For instance, could he safely interact with people in the past? He wouldn't want to create some kind of butterfly effect to negate his own future existence.

Angus's findings were illuminating. Some PhD guys had identified a self-consistency concerning the past. Namely, that it protects itself. Angus could enjoy minor interactions, such as inconsequential conversations, during visits to bygone days. Such would have no effect on the future. But he couldn't attempt substantial change. Say, like the cliché of killing his grandfather. Then he'd be in danger of creating a time paradox. If he got too close to the line, physical laws would prevent such deeds. Much like a concrete wall would stop someone from walking through a building. Except in one scenario—if events had originally transpired

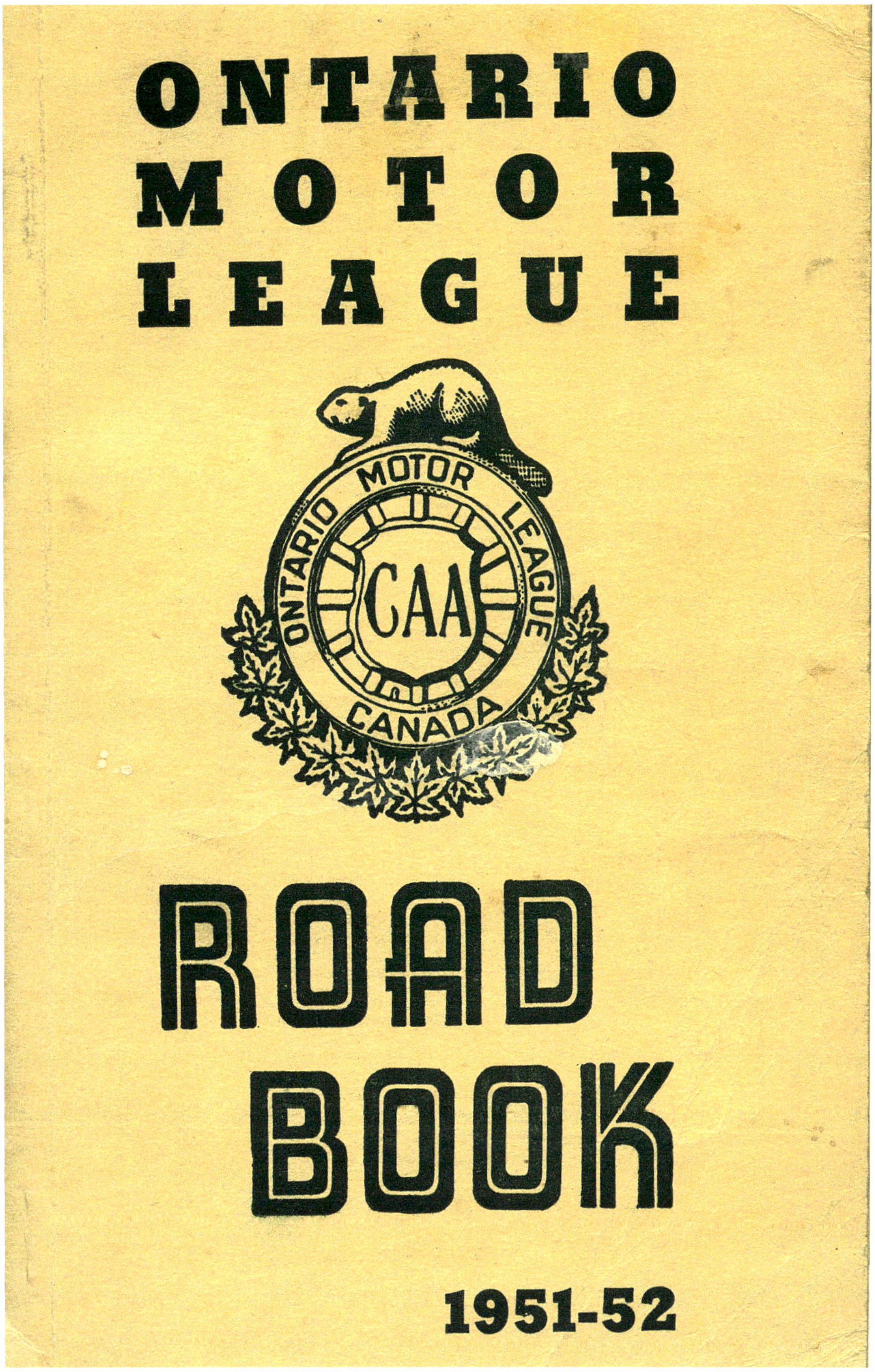

differently from what he'd understood. In that case, he would appear to "alter" the past to fulfill his role in history. In other words, he'd been there all along. He just wouldn't have known it, in his present existence. His role in the past would be a self-fulfilling prophecy.

After visiting Kitchener and Bobcaygeon, Angus convinced Eddie to send him on a more ambitious trip. He wanted to cover a wider swath of territory and stay longer. Eddie couldn't be sure what controlled the duration of anyone's stay in the past. By process of elimination, he figured it was the battery charge.

"She don't lose juice when the motor's running. With the engine off, it depletes. You know? When she runs down enough, you're back to the present." Eddie tapped his temple. "Near's I can figure, anyways."

Angus understood that. It was similar to when a smoke detector ran down and started beeping and flashing. The reversion to the present held for car and occupants. Looking at the calendar for July 1957, he saw a whole stretch of dates with readings. He wanted to stay in the past for several hours on those days. Enough time to drive around and find lots of operating steam locomotives. For that, he would need 1957-era provisions—a map, money, clothing, identification. Some automobile know-how, too. Over the course of a week, he made those preparations with Eddie's help.

Amanda was skittish about going back to 1957 again. She resisted Angus's efforts to cajole her. Even getting her to climb into Eddie's restored car was out of the question. Angus had known her long enough to understand that was par for the course. As Eddie said, "it takes awhile to get a woman to come around to your way of thinking." She finally relented, on one condition—that Angus do another trial outing. He arrived at the old White Rose station late afternoon. That was just yesterday—a day before his visit to St. Marys.

Eddie had wires from the motor analyzer hooked up to the battery in the '56 Ford. "Doing some amperage and volt readings on the engine. Gotta make sure my conjecture's right." Angus and Amanda hung back in the shadows, shivering and watching. It had been dark for two hours when Eddie finally instructed Angus to get behind the wheel.

"Start 'er up." Eddie squinted at an entry on his 1957 calendar. "If I'm right about the battery, this'll only last a few minutes." He nudged Amanda with his elbow and winked. "You'll be able to tuck him in with his cookies, just like any other night." She rolled her eyes.

Angus revved the motor. He felt like an astronaut must, just going up into orbit instead of to the moon. The last thing he remembered in the present was looking out at Eddie. He was pointing at a pencilled scrawl on the calendar for Amanda's benefit. Then the frigid March darkness of Washago transitioned, in a moment, to a summer night.

There was a building now in front of the '56 Ford. It loomed in the darkness, blacker than the night sky. An open door in its near wall seemed to beckon Angus. He would act on such signs, trusting the prompts to send him in the right direction. Mindful of being on a trial run, he turned off the engine and climbed out of the car. There was a heavy, muggy feeling to the night air. He approached the doorway, then stepped across its threshold.

Inside the building were steam locomotives—more than Angus had seen in his life. He'd entered the sultry gloom of a railway roundhouse. How amazing was that? Light bulbs cast a yellow pallor over the surroundings. The aroma of coal smoke permeated the moist air. Probably a dozen engines in all rested on tracks that radiated out from a turntable. He walked around the outside face of the structure, past the noses of locomotives. His shoes crunched on cinders which coated the concrete floor. There was a crew despatch chalkboard on a wall. He moved closer for a closer look. It turned out he was at Sarnia. That was southwest of London. The date was July 22, 1957. Way to go, Eddie, he thought. You've done it again.

Angus turned back to the locomotives. The nearest was number 6232, of the 4-8-4 wheel arrangement. She was a Northern Type, and would be scrapped long before his time. There were several more sister engines in other stalls. Most were numbered in the 6300-series, but there were a couple of 6100s too. It was unbelievable being in the presence of so many live steam engines. They were massive compared with the Ten Wheelers Angus had encountered at Bobcaygeon.

There was one small locomotive which caught Angus's eye—number 48. As he faced engine 6232, she stood to the left. She would have fitted in with his Thomas the Tank Engine set from childhood. This resident of the

—Jim Shaughnessy photo

Sarnia roundhouse, July 22, 1957.

—John Rehor photo

roundhouse had no tender. Just then, a shop man, clutching a clipboard, emerged from the shadows. Angus shuddered and his heart raced. Fighting to remain nonchalant, he nodded at the subject of his fascination without saying anything.

"Her?" The shop man rested his hand on the engine's pilot beam. "She's going out on 606 for London in the morning."

"Oh?"

"Yep. Might be her last trip from here, though. Back to Montreal any day."

Feeling his unease draining away, Angus turned to face the man. "You don't have any use for her?"

The shop man shrugged. "All she's good for are London turns. Even then, we normally only use her Saturdays."

"How'd she end up here?"

"Montreal cut back on commuter runs couple years back. So, they sent her to Allandale. Only Allandale cancelled the branchline train she was meant for. Then she sat idle. Better part of eight months in their roundhouse. Till our Weasel broke down in December of '55. Would take months to repair. So we got the tank engine."

"What's this thing you call *Weasel?*"

London,
July 16, 1955.
—author's collection

"Our diesel car, number 15836." The shop man chortled. "Basically a green varmint that slithers atween here and London."

Angus laughed.

"We've got other names for it." The shop man showed a row of uneven teeth. "*Contraption. Wreck*. Thought we'd pawned it off on Allandale this spring. But they sent it back. Now the damn thing's laid up in London. We been using road diesels and 5500s on the turns." He nodded at engine 48. "I'll say one thing for this girl. She's fast. Up to 65 mile per hour lickety-split. Hopefully, Montreal can still use her. Otherwise, it's the scrap heap."

The shop man took a pencil from behind his ear. He scratched a note on a *Locomotives in Terminals* report, then walked away. Angus watched him pull a piece of chalk out of his pocket. The shop man approached a 6300-class engine and scrawled something on its smokebox.

Angus contemplated what had been transpiring. There he was, on railway property. In a roundhouse, to boot, with no one bothering him. The shop man assumed he could take care of himself. He wasn't paranoid about Angus being injured. No one was demanding to know what he was doing there, insisting he wear a hardhat and safety boots. It was a remarkable contrast to the world of his present day.

The shop man's proclamation of doom for tank engine 48 still rang in Angus's ears. He stood close to the locomotive, listening to its creaks and groans, watching wisps of escaping steam. The engine's dynamo, an electric generator on top of the boiler, whined. Angus cocked his head at the sound. It got louder, turning into a screech. In front of his eyes, engine 48 melted away. Everything went white. Then Angus was back at Washago. Eddie and Amanda were standing on the driveway.

That had been last night. This morning, Eddie had foreshadowed what was to come. "You're heading for Stratford when Amanda gets here." He'd asked Angus when she would arrive.

"Well, she's just started walking the kilometre from her house to your garage."

"Okay, then. Want to take a quick look at St. Marys?"

"Sure."

Now, Angus has had enough of watching men load express packages aboard train 168 from a distance. He doffs his winter jacket and tosses it into the back seat.

As Angus's feet touch the 1957 pavement, he feels the familiar sense of unease. It concerns him standing out from everyone else. Not because of his clothing or haircut—after the initial visits to 1957, he's donned appropriate attire and groomed himself accordingly. No, the apprehension relates to his lack of purpose at the Canadian National Railways' station. Unlike the other people, he has nothing to do. It's akin to the isolation he's felt on social occasions. Not knowing anyone present, but struggling to justify standing alone by choice. How much scrutiny will he face during these visits to the early Cold War era?

Overriding his unease, Angus approaches the station platform. Even at quarter to eight—he spies the time on an express clerk's wristwatch—the day is heating up. Now he's right up beside the green and black train.

Over Angus's left shoulder, the station door opens. Two women around his age—early twenties—emerge onto the asphalt platform. They approach the passenger coach. Angus admires their forms in floral print dresses and sun hats. Their white high-heeled shoes match their cotton gloves. Red lipstick sets their mouths aflame. They have trim figures, perfect posture, and elegance. Too good to be true, as if clipped from a 1957 Eaton's catalogue? No, they are real. They are a tonic to eyes accustomed to the anything-goes norm of the 21st century.

The first woman is carrying her gloves instead of wearing them. Angus discovers why. She swivels her left wrist to display a diamond ring to her companion. In reply, the second woman laughs. It is a bright, cheerful sound.

Has Angus encountered that second woman's likeness in his present world? Yes, at least once. He carries a picture of her in his wallet. The woman takes the hand of

For train 169 at St. Marys, January 1959.

—James A. Brown photo

Train 168 at St. Marys, July 23, 1957.

—Don Wood photo

the conductor, who is standing beside a portable step. She follows her friend up the vestibule. Angus hears her voice through the open coach windows as she finds her seat. It has a pleasant ring, like a church bell on Sunday morning.

"Engaged to John Hanlan? Good heavens, poor John. When I turned him down, he warned me he'd do something desperate."

Startled out of his self-conscious funk, Angus guffaws. A couple of Royal Mail men cast scowls his way.

Angus looks around. Everyone is engaged, doing something useful. But they have eyes and ears. How many are appraising him? For their sake, he'd better figure out what he's doing on the St. Marys station platform. He wants to walk forward alongside the train to check out locomotive 5594. But he'd better come up with a story before conversing with the engine crew. What could it be? He seizes on an idea. The smart-aleck woman is his girlfriend. Wouldn't that be nice, in 1957? He's dropped her and her companion off so they can take the morning train to Stratford. They're going to meet a dressmaker lady—that's it—who designs bridal wear. She's a seamstress who otherwise makes costumes for Stratford's Shakespearean Festival.

Mid-step toward the engine, Angus notices a man staring at him from inside the depot's bay window. The fellow's rolled-up-sleeves speak of a businesslike demeanour. He looks very much in charge. Say, in the capacity of a CNR station agent? He's probably sitting where he does every morning, with a view up and down the tracks. But today the train is not commanding his attention. Instead, he's scrutinizing Angus head to foot. And now he's getting out of his chair and coming outside.

No point in scurrying away, Angus figures. He was hoping to first test his alibi with an ordinary passerby. Maybe even Gladys, through the coach window. He feels her gaze on him too. But starting with that authority figure from the station is another story. Angus's brain commands his feet to walk toward the '56 Ford, and pronto. But his shoes seem cast in concrete. He voices a wish in his head. Come on, car battery. Run out. Make that buzzing sound start. Get me back to my regular life. I'll give St. Marys another try after Eddie recharges the juice.

It wasn't working. The man from the station moves beside Angus, nodding and smiling. In a probing fashion, it appears. He speaks in a clipped tone.

"Gladys is a pretty one, all right. Are you her new beau?"

St. Marys, April 13, 1957.
—author's collection

Angus's jaw drops open. He looks at the man, not sure how to answer. Hookah Hopkin, the railway history buff, has told Angus about the steam era. One thing he talks about is the railway station agent. In steam days, he's the centre of his community. It's his business to know everything and everyone. That includes Gladys, whose chin rests on her hands as she stares out the coach window. The station man doesn't know Angus, a stranger with an imaginary purpose. But it's his job to divine what that purpose is. Here goes, Angus says to himself. Let's hope the guy's friendly like the Sarnia roundhouse man.

"You know," Angus hears himself saying as he waves at Gladys in her coach seat, who returns the greeting with a smile and batting of her eyelashes, "some days I wonder." It's dull and cliché, but it delivers a satisfying completion to the exchange. He laughs with the station man. Would there be a follow-up question?

From the head end of the three-car train, locomotive 5594 whistles twice. There is a hiss as the air brakes release. Angus turns away from the station man and watches the engine. Puffs of smoke rise from the stack. Clouds of steam blast from the cylinders. The drive wheels begin to revolve, creaking the passenger cars into motion. Train 168, with Gladys and her friend aboard, is on its way to Stratford.

"They had that engine on the Allandale run out of Hamilton a month or so back." The station man must have noticed Angus's fascination with the locomotive. "Took over from the Weasel up there."

"Yes." Angus is delighted to be able to converse with knowledge. "Diesel car 15836 from Sarnia." The station man nods. He's the second person in as many days to mention the place called Allandale. It's in the south end of Barrie, Angus knows. The old railway part of town. About half an hour's drive from where he lives. Maybe Eddie can send him there someday.

The coach carrying Gladys passes Angus's field of vision. He waves to her, partly to affect an air of familiarity for the station man's sake. She tilts her head and offers Angus a mischievous smile. Then she taps her fingers against the window ledge, as if playing trills on a piano. She locks her eyes on his. Angus wonders what will become of her over the years? Would she still be alive in his present day? He turns back to the station man.

"Are you the agent here, sir?"

"Me?" The man laughs. "Nope. Just a ticket clerk. The agent doesn't come on till noon. I started at six. That's railway time, you know." He looks at Angus. "It reads an hour earlier than your watch, if you were wearing one."

"Right." Angus taps his wrist, thinking of a story for the watch that should be there.

"Supposed to be cooler today." The ticket clerk looks up. "Those thunderstorms yesterday did the trick, yes sir." He stands beside Angus, arms crossed. He watches the train's last car, an express refrigerator carrying marker lights, move up the tracks. "Well, I'd better put number 168 on the sheet for the DS in Stratford."

Thanks to Hookah, Angus knows what DS means—despatcher. The ticket clerk raises a hand to one side of his forehead in an informal salute. Then he walks through the doorway and back to his desk in the bay window. The Royal Mail and CN Express men depart like the train. Their trucks rumble away, past the spot where the '56 Ford is parked.

Angus is alone again. I probably won't be long in St. Marys, he thinks, but Stratford is coming up. That's welcome news about the heat wave having passed. It already feels like 80 degrees just the same. The weather will be a good conversation starter when Amanda and I visit Stratford today. But what if wisecracking Gladys in the floral print dress really does get off the train there? Who knows what she might say. It could get awkward. Oh well, Gladys and I have no future together. If I impact her life at all, it'll only be in a glancing manner. Right. Unless I've been here all along. She sure reminds me of Amanda—and even looks like her.

What's struck Angus about St. Marys is how routinely people are carrying out their roles. But why shouldn't they? It's he who is the alien in their world. They're just going about their normal lives. Part of him wants to jump into their midst and tug their sleeves. Maybe gather them around in a circle. Explain that the events of Tuesday, July 23, 1957 are fleeting. They'll fade away like the chalk marking the date on the schedule board affixed to the station. Poof! All of this will disappear into the pages of time. Gone forever. But they know that already. And they'd label him a whacko, with good reason.

It's just me longing for a past I've never known, isn't it? Angus muses. No one at St. Marys is enamoured with things like I am. Such as the Canadian National Exhibition sign leaning against the front of the station. It's identical to the one at Bobcaygeon, listing the Grandstand acts from August 23 to September 7. And how many people here are aware that the Age of Steam is about to end? Does it matter to them? It might to the men who work on locomotives.

St. Marys, July 21, 1958.
—Howard Ameling photo

But to Gladys and her friend? They just want to ride to Stratford, say, to see a dress designer. Obviously, they'll prefer diesel-hauled air-conditioned coaches to the open-windowed variety positioned behind soot-belching iron horses.

The adrenal rush associated with landing in an unfolding scene of a bygone era is growing on Angus. This has been nice, he reflects, visiting St. Marys on July 23, 1957. With Eddie's time machine, I can escape my present-day world for the past whenever I want. It's a liberating feeling, like having an ultimate lifestyle choice. It will be important to register my impressions of the things and people I discover. These trips to the past are demanding to be recorded and shared with others. Maybe Old Eddie knows that too—which could be his purpose in the big picture. Angus chuckles to himself. In a way, he's like Steinbeck or Hemingway or Twain—men who used real-life experiences for stories. In Angus's case, Eddie is supplying him and Amanda with visits to the past as raw material.

Angus peeks through the bay window at the ticket clerk, who is doing paperwork. He hesitates for a moment. Dare I ask him about trains? Yes. Angus walks to the door leading into

Way freight at St. Marys, July 27, 1958.

—Jim Buckley photo, Al Chione collection

the station office and twists the knob. Nosing inside, he speaks loud enough to attract the man's attention.

"Do you have any more steam engines coming through today?"

"Steam engines?" The ticket clerk stares at Angus, bug-eyed. "That's all we ever have coming through here."

A fortunate place to watch trains, Angus thinks.

"Mind you," the ticket clerk goes on, "the regular operator's been getting worried. Figures steam's on the way out."

He's right, Angus almost says out loud.

The ticket clerk reaches for a clipboard hanging from a hook. "Let's see what the lineup says." He focuses his eyes on the top page. "Not much, as usual, after number 168. The way freight from Stratford will be along. Probably around ten this morning."

"What kind of power on the way freight?"

"You're interested in this kind of thing, son?"

Angus nods.

"More 'n likely it'll be engine 2629. She's what they call a Consolidation, a 2-8-0."

"Yes." And I wish they'd saved more of them from scrap, Angus wants to say.

"They've been busy at the cement works lately."

Angus raises his eyebrows, inquiring.

"Just a mile down the track." The ticket clerk points off to his right, toward London. "I'll tell you something." He lowers his voice. "The owner of St. Marys Cement has a lot of political pull. That's why every passenger train stops here. If they didn't, he'd give the railway business to the CPR. And you know something else?"

Angus leans forward, cocking his head.

"If the cement company tells the CNR they won't have their outbound loads ready till three or four in the afternoon"—the ticket agent swats away a buzzing house fly—"then the way freight crew will have to twiddle their thumbs until then."

Angus makes a mental note to watch the way freight switching cars at the cement company someday.

"They'll still be switching cars there when 29 comes through from Stratford. About four hours from now. That's one train you see the occasional diesel road switcher on. Whenever they put an assist engine on the *Dog* out of London, it deadheads back on 29. Normal power for 29 is light steam."

"What's this *Dog*?"

"Night time short haul freight, Stratford to Stratford via London. Came back through here on the return trip an hour ago." The ticket clerk looks back at the list. "Oh, here's something for you. Apparently there's a light locomotive coming through, early afternoon. It'll be Extra 7427 West."

"That's a yard engine, six-coupled." Angus remembers seeing one running. That was—or will be—down in New Hampshire on a tourist railway.

"Yeah." The ticket clerk looks Angus up and down. "Where's your camera?"

"Huh?"

"You train buffs always have cameras slung around your necks." The ticket clerk taps the paper on the clipboard. "You'll want it for this one. Engine 7427 will be all shiny. Just out of the Big Shop."

Margaret's Irish Peer Beau Sent To Cyprus

METRO EDITION

INDEX
Amusements—8-9
Births, Deaths—22
Bridge—10
Comics—37
Crossword—18
Horoscope—26
Markets—16-17
Racing—14
Radio, TV—22
Shipping—2
Sports—12-15
Suburban—7
Want Ads—27-36
Women's—23-26

TORONTO DAILY STAR

THE WEATHER
Toronto and vicinity — Tuesday: Cooler. Sunny. Low tonight, 60; high Tuesday, 80.

65TH YEAR — Authorized as second-class mail, Post-office Department, Ottawa. — MONDAY, JULY 22, 1957—40 PAGES — 5c Per Copy — 35c PER WEEK HOME DELIVERED

Miss Toronto of 1957

—Star Photo by Federal Newsphotos

PRETTY MARIANNE LENCHAK, blonde, hazel-eyed and 20, was declared Miss Toronto of 1957 at the annual police games Saturday, as she defeated 50 other contestants. Oshawa-born Miss Lenchak was given an ovation by spectotors

GIRL, 15, DRUNK ON VODKA

GIRL, 7, BOY, 5, DIE
7 FLEE GALT FIRE

SEVEN ESCAPE HOME ORDERED TO BE REWIRED

Special to The Star

Galt, July 22—A boy and his visiting girl cousin were burned to death and seven other persons, including three children, narrowly escaped in their night clothing when fire gutted a one and a half-storey frame house early today.

The bodies of Grant Hastie, five-year-old son of Mr. and Mrs. Kenneth Hastie, and Diane Hastie, seven, of Dundas were recovered by Police Constable Ed Guthrie from the charred front bedroom beds. They died despite efforts of the boy's father to reach them.

Kenneth Hastie suffered burns and cuts about his face, arms and body in rescuing his three other children. He

GET LEG IN LAKE MAN MURDERED BODY CUT UP?

Special to The Star

Montreal, July 22—Discovery of a man's leg has led senior police officials to believe Larry Petrov, Montreal welder and friend of George Lemay, has been murdered and his body dismembered.

Dr. Rosario Fontaine, provincial medico - legal expert, established that the leg came from the body of a "heavy-set, brown-haired man." Petrov was five feet, 11 inches tall, weighed 225 pounds and had brown hair and brown eyes.

An attempt will be made today, police said, to establish positive identification through a sock found on the decomposed limb. The limb was so swollen it was first believed to belong to a woman, police said after it had been reported found in Lake Ouareau about 80 miles north of Montreal.

Petrov disappeared July 2 but the fact was not reported by his parents until a week later. When police began their check they questioned Lemay

Would Angus see that engine at Stratford later this morning, getting ready for transfer to London? He'll keep an eye open.

"After that, nothing till the way freight goes back through toward suppertime." The ticket agent looks at a wall clock. "Gotta go, son. Need to do a yard check at the junction." He points to his left, toward Stratford, then gets to his feet. "Anyways, tonight. The agent will have four passenger trains through in a couple of hours. Numbers 171, 170, 11 and 40. All behind steam, far as I know. Then the *Dog*. Usually a Northern fronting that one."

Angus shakes hands with the ticket clerk. "Thanks so much, sir." He walks along the platform, then away from the station toward the '56 Ford. There is a man a few paces ahead of him. He's carrying a folded newspaper in his hand. Angus gets closer and angles his head. It is yesterday's Toronto Daily Star. The man pauses at a garbage receptacle. Just as he's about to stuff the newspaper in, he notices Angus's interest.

"Here." With a smile, the man hands over his newspaper.

"Thanks."

Back in the front seat of the car, Angus unfolds the July 22, 1957 edition. He scans the headlines. If the front page is any indication, it's a brutal world he's visiting. Murders and executions. Accidental deaths—fires, drownings, automobile crashes. He remembers the words of his driving instructor—there's no such thing as an accident, everything happens due to a logical chain of events. Looking over the accounts of tragedies, Angus realize he's right. All the deaths, some 56 of them across Canada on the past weekend alone, are preventable. Or were. His advice to the people of 1957 would be simple. Don't drink and drive. Wear seatbelts and life jackets. Use the gadgets called "fire detectors" already being advertised in publications.

Saving lives is beyond his remit for 1957, he knows. But that might not stop him from trying. What's the worst that could happen?

There's Miss Toronto on the front page of the Daily Star. A newsprint headshot, life-sized, smiling at Angus. *Blonde, hazel-eyed and twenty.* She's pretty. Like Gladys and her friend—and Amanda. She's the only joyful subject chosen by yesterday's front page editor. The other wire service accounts unnerve Angus. Children perishing in fires. A dismembered body found in a lake. Hangings. Boating accidents. Snake bites. A few years on from 1957, Marshall McLuhan will make his famous assertion. He'll say that this newspaper—which chooses to chronicle death and suffering—is itself the message. These stories arrayed across the broadsheet? They are a reflection of 1957 readers.

St. Marys, July 21, 1958.
—Howard Ameling photo

From the front seat of the '56 Ford, Angus gazes along James Street. He watches the trail of smoke from locomotive 5594 dissolve in the morning haze. The train has just traversed a long steel girder bridge. It rests on limestone piers and spans a valley. The watercourse underneath must be Trout Creek. That is according to a sign attached to a nearby streetlight. Its arrow points to a park in that direction. A nice place to sit with Amanda, watching trains, maybe do a little fishing. Angus watches the green and black cars of the train disappear behind some trees. He wiggles the key in the ignition, leans his head back against the seat, and waits. When will the mechanism whisk him back to his regular life? Eddie said this would only be a warm-up, passing time before Amanda arrives.

Then, something intangible happens. For Angus, it's a premonition. Like one he feels on a midway ride at a fall fair. It occurs in that moment just before the tattoo-bedecked operator decides the patrons have had enough amusement. Angus knows the ride is about to end. Now he senses a change in the air around him. Right about now, the Zipper or Wild Mouse would let out a hiss, then decelerate. What's happening at this moment, parked on a side street in St. Marys? Is it a pressure change? High-frequency vibrations? Something atmospheric, for sure. In the blink of an eye, Angus is back to Washago. Around him is melting snow, plus the junk which clutters the yard of Eddie's former service station. He climbs out of the Ford Mainline.

Eddie is standing on the driveway, as he was before. "How long were you in St. Marys for?"

"Oh"—Angus shrugs his shoulders—"no more than fifteen minutes I'd say."

"Good." Eddie hoists the hood and starts hooking a recharger to the car's battery.

Angus walks onto the abandoned stretch of Highway 11 fronting the old service station. He looks northward, to his right. Amanda will appear any minute on foot. She'll be taking the

—author photo

short cut this side of the CNR tracks. Unless she's lost her nerve, that is. But no, there she is, wearing a knee-length coat over bare legs. She's planning ahead. Angus waves to her and she responds in kind. Now she's running her gloved fingers along the metal railing of a short bridge. Pausing to look down at the water in the creek. That's not like her. She's not one to dawdle. Maybe she's trying to anchor herself in today's world.

So, it will be Stratford for Angus and Amanda, on July 23, 1957. He can hardly wait. Hookah Hopkin told him it was the last good year for steam locomotives in Ontario. The CNR backshop in the Festival City—as it had come to be known to people uninformed as to its railway history—turned out a record number of rebuilt engines that year. Then it all went downhill for steam.

There's something Angus has agonized over, ever since he learned about the time-travelling 1956 Ford Mainline. Eddie is way past 80 years of age—no one really knows how far. The matter of his dying is a ticklish thing to bring up with him. But what if he passes away suddenly? The knowledge of the time-travel process would go with him. Angus would lose his portal to the past. He and at least one other person—Hookah perhaps—had better learn the secrets. They'd need Eddie to train their ears to detect pitch and rpms of the Ford Mainline engine. What else did Eddie know about it, though? Were there things he wasn't telling them?

Amanda offers fist bumps to Angus with glove-clad hands. "Props."

"You nervous?"

"Yes." Amanda flips the straps of her knapsack, which is bulging, off her shoulders. It would be packed with extra summer clothes and other goodies they discussed last night.

"That's good."

"That I'm nervous?" Amanda narrows her eyes at Angus. "Why?"

"Well, it's not like you to be blasé about going back in time. So you must be ready." Angus flips his head toward the '56 Ford. "My stuff's already inside."

Stratford, September 29, 1957.
—Alan Crompton photo

Eddie disconnects the leads from the car battery and slams the hood shut. With a look at Amanda, he walks around to the passenger door. He grabs a lever and angles the seat forward. Amanda follows him. She stashes her knapsack alongside Angus's gym bag on the back seat.

Amanda settles into to the front seat on the passenger's side. Angus takes his place behind the wheel.

Eddie leans his head through the open driver's door. Up close, his face has the leather texture of a reptile's skin. "You've got a full tank of premium gas. Hard to come by nowadays. It'll take you about 240 miles. Where you're going, fuel sells for 22 cents a gallon or thereabouts. Make sure you fill 'er up." He nods to Amanda. "Got everything you need in the parcel compartment?"

Amanda scrunches her face, looking puzzled. Then it hits her. She reaches for the latch of the glove compartment and opens it. Inside is a stack of items, which she pulls out. Angus already had some of them out on the seat at St. Marys.

"A 1957 Imperial Oil road map. It's the one that shows railways." Amanda flicks a smile at Angus. Her overall manner is terse, though. She's radiating a sombre calm, like crew members before a rocket blast-off at T-minus 90 seconds. Launch Control: Pressurized first stage LOX tanks to relief. Astronaut: Pressurized. Amanda flips through the rest of the documents. "Railway time tables, CNR and CPR. A 1951-52 Ontario Motor League road book." She pouts at Eddie, with the sunlight sparkling off her eyes. "It's five years out of date. Is that the best you can do?"

Eddie shrugs. "Take it or leave it." All three of them laugh. Eddie turns to the driver. "Got the most important thing?"

Angus taps the front pocket of his casual slacks. "A wallet full of 1950s cash."

"Don't leave home without it." Eddie clears his throat and spits on the melting ice on the pavement outside the service bay.

Some 20 years ago, when the railway steam era in Ontario was still strong in living memory, I set out to create a snapshot of that world. Over the course of a decade and a half, I created a series of nine hardcover books. When you read one, you are taken back to a day I chose—Friday, June 25, 1954—to experience the sensory aspects of that wondrous age. Now, *King's Highways & Steam Trains* does that too, with added twists.

That aforementioned award-winning series could not be produced today. The binders of notes I accumulated, from interviewing men who worked in the steam era, now represent memories and impressions of those passed. But for those handwritten transcriptions, the day-to-day intricacies of a bygone era would be lost to time.

Over the course of the two decades since I began my landmark series, the demographics of my reading audience have changed. Where once I sought to remind the typical reader of an era he had once known, now I seek to convey to people of later generations an account of an era—and possibly locale—with which they have no firsthand association. Thus, *King's Highways & Steam Trains*. In this series, projected to number at least half a dozen volumes, I widen the lens through which I view and portray the past.

So, meet Angus Wolfe and Amanda Webb! These characters were introduced as children in my debut novel entitled *The Secret of the Old Swing Bridge*. As 16-year-olds, the homeschooled heroes carry on their amateur investigations in *The King's Puzzle* (as of this writing, four of six novels in the serial have been published in eBook format). Angus and Amanda are the point-of-view characters through which my 21st century audience experiences the past. It's not only railway operations that *King's Highways & Steam Trains* portrays. As the title suggests, each of these new volumes will offer a glimpse at a brief era—from the end of the Second World War through the 1950s—when the highways and railways of Ontario coexisted as complementary transportation systems.

As readers of *Steam at Washago* will know, I lost my parents over the interval from 2011 to 2015. Having stepped back and glimpsed their presence in my life through a wider perspective, a symbol has emerged for me. That is a certain two-door sedan, a 1956 Ford Mainline in Bermuda Blue. For the first two years of my life—the impressions of which formed the basis of my life work—I viewed the larger world from the front seat of that automobile.

Though the memories live only in a recess of my subconscious, I experienced the vestiges of the world described in *King's Highways & Steam Trains* from my perch between my parents in the front seat of that '56 Ford: Grass-grown railway branchline rights-of-way which still sported red depots frequented by passenger trains. Lines of condemned steam locomotives at Allandale. A CNR Northern despatched along every railway line in the southern part of the province which afforded a place to turn an engine. The entire milieu of that era embodied by the phrase "down by the tracks". It was still there in the early 1960s—or almost still there.

Though I came along just a little late to experience the era portrayed in *King's Highways & Steam Trains*, my dad's 1956 Ford arrived in plenty of time. That car explored a myriad of railway mainlines and branchlines in Ontario by way of the King's Highways and back roads.

The 1956 Ford of this book series is a time-travel vehicle. For me, that is not such a leap. The real car on which Eddie's vehicle is based—the one my dad bought new from an Ajax dealership—is also a time travel vehicle. It is the portal through which the stories of these volumes are delivered. Like the editor of the *New York Sun* explaining Santa Claus to an inquiring Virginia, I tell you that Dad's 1956 Ford is a time-travel vehicle! It is a physical and ethereal portal to the adventures of Angus and Amanda in July 1957.

Many people have assisted in the creation of this first volume of *King's Highways & Steam Trains*. I would like to name them here.

Jim Brown, Al Chione, Keith Hopkin, the late John Rehor, Jim Shaughnessy and the late Don Wood assisted with railway photographs from the 1950s.

Research, analysis, reminiscences and supporting documents were provided by Jim Brown, Brian Davis, Keith Hopkin, Ron Johnson, Gord Launchbury and Jeffrey Smith.

My email list, called the "Branchliners", was instrumental in the creation of this debut volume. By poll, they chose the final title from a list of several. They offered critiques of various cover designs, and chose the ultimate candidate by vote. They also provided reminiscences of the 1950s. Thank you, Branchliners. If you haven't already joined their number, kindly visit **www.ianwilsonauthor.com** and sign up for news of releases and other goodies.

Sixteen of those Branchliners helped in a more substantive way, by reviewing an early draft of the manuscript for the first volume of *King's Highways & Steam Trains*. For that, I would like to thank (in alphabetical order) Pete Beggs, John Bigham, Doug Currie, Brian Davis, Doug Gifford, Harper Habbersett, Dave Huskinson, David Jennison, Bill Jolitz, John Martin, Doug Matheson, Mark Michener, Andy Panko, Sharon Perry, David Shearman, Paul Smith, Malcolm Vant, Hannah Wheelband, Ian Wheelband and Bruce Wilson.

Lastly, and most importantly, I would like to thank my wife Mary-Jo and teenaged sons Spencer and Duncan. They watched this project incubate from an idea to a handwritten draft to a finished document. Along every step of the way—brainstorming story ideas, scanning and archival photography, proofreading, promotional videos, mailing list management—they've helped with enthusiasm. As you read this book in your hands, also know that they carefully packaged and shipped it!

Orillia, Ontario, October 2, 2016